150 YEARS OF PURDUE ENGINEERING

Celebrating Our Consequential Impact on the World

ARVIND RAMAN

JIM SMALL

Purdue University Press
West Lafayette, Indiana

Cataloging-in-Publication Data on file at the Library of Congress.

978-1-62671-292-8 (hardcover)
978-1-62671-294-2 (epdf)

Cover image: Courtesy Paul Sadler with the ESC Plan.

CONTENTS

VII **PREFACE**
A Special Message From Dean Arvind Raman

1 **CONSEQUENTIAL STORIES**
150 Consequential Stories From Purdue's College of Engineering

83 **CONSEQUENTIAL LEADERS**
The Deans Who Have Guided Purdue Engineering for 150 years

87 **CONSEQUENTIAL INDIVIDUALS**
Purdue Engineering Alumni and Faculty Elected as National Academy of Engineering Members

95 **INDEX**

97 **CREDITS**

Purdue University Sep. 15 1874

Faculty met. Present A.C. Shortridge, J.S. Hougham W.B. Morgan, J. Hussey, H.W. Wiley E.F. Brown.

The following Special Course in Engineering was adopted as reported by W.B. Morgan.

Course in Engineering

The courses of study in the several departments of Engineering are for the present identical as far as it seems practicable to make them. During the Senior Year the Students can pursue as a specialty either Mechanical, civil or mining Engineering, or Architecture. In the Freshman Class the studies are the same in all the Schools of the University.

Minutes from the Purdue University faculty meeting held Sept. 15, 1874, outline the establishment of new courses for the study of engineering. Students could specialize in mechanical, civil or mining engineering.

PREFACE

As we commemorate "150 Years of Purdue Engineering," we are celebrating myriad heroes—remarkable Boilermakers whose breakthroughs have fueled our journey to become the nation's most consequential engineering college.

This book highlights the indomitable spirit, ingenuity and achievements of 150 of those alumni and faculty members whose innovations have improved nearly all facets of society. Accomplishments have ranged from pioneering fundamental technologies to medical innovations to space exploration.

Among examples, you will read how Boilermakers invented the most manufactured item in history, barcode, wiki, electronic TV receiver, soft-serve ice cream machine and artificial kidney. The book also showcases the first female solo transatlantic flyer, first and most recent moon walkers, first female commercial astronaut, "First Lady of Engineering" and "Father of Radio Broadcasting."

Building on this rich legacy, Purdue Engineering has never been larger or better. We keep persevering to expand our impact in incubating talent, increasing research influence, and advancing the state of Indiana and the nation. We invite you to take pride in our first century and a half, paving the way for our next 150 years.

Arvind Raman

ARVIND RAMAN
John A. Edwardson Dean of the College of Engineering
Robert V. Adams Professor in Mechanical Engineering
Purdue University

CONSEQUENTIAL STORIES

150 Consequential Stories From Purdue's College of Engineering

U.S. PARACHUTE TEAM

ENGINEER, U.S. ARMY OFFICER, MEDICAL DOCTOR, JUDGE, AND SKYDIVER

Sue Hudson Abreu received a bachelor's degree in interdisciplinary engineering (biomedical) in 1978. Following her time in Army ROTC at Purdue, she embarked on a military career and earned an MD at the Uniformed Services University of the Health Sciences, and specialized in nuclear medicine. Abreu served in various capacities during her 24-year U.S. Army career, including nuclear medicine consultant to the Army Surgeon General, and retired as a colonel in 2002.

She has served as president of the American College of Nuclear Physicians and president of the Nuclear/PET Division at the Intersocietal Accreditation Commission. Abreu also earned a JD from Oklahoma City University's School of Law in 2011, which led to her 2015 appointment as an administrative judge (technical) at the U.S. Nuclear Regulatory Commission's Atomic Safety and Licensing Board Panel. She became the associate chief administrative judge (technical) in 2019.

Beyond her professional endeavors, Abreu is an experienced skydiver who has served as team manager and team physician for the U.S. Parachute Team for many years. In 2004, she was honored as a Purdue Old Master and inducted into the Purdue Reserve Officers' Training Corps (ROTC) Tri-Service Hall of Fame.

Purdue presented her with the Outstanding Interdisciplinary Engineer Award in 2001, the Distinguished Engineering Alumna Award in 2020, and the Engineering Alumni Loyalty Award in 2023.

EVP AND CFO OF THE CAMPBELL'S COMPANY

Carrie (Byrnes) Anderson earned her bachelor's degree in chemical engineering in 1991. As a student, she interned at General Motors and secured a position with GM immediately after graduating. While at GM, she earned an MBA. In 1998, she transitioned to an investor relations role in GM's Treasury Office. A year later, she joined Delphi Corporation, where she quickly rose to vice president of finance, and divisional chief financial officer within six years.

In 2011, Anderson moved to Dover Printing and Identification, where she served as vice president and CFO. In 2017, she was promoted to chief accounting officer and corporate comptroller for Dover Corporation. Two years later, Anderson assumed the role of CFO at Integra Life Sciences, where she served until 2023 when she was appointed executive vice president and CFO at The Campbell's Company.

With more than 19,000 employees, Campbell's is one of the largest food companies in the U.S., boasting iconic brands such as Campbell's Soup, Pepperidge Farm, Snyder's of Hanover, V8, Pace Foods and Swanson. Among many honors, Anderson has been recognized on Institutional Investor's 2022 All-America Executive Team, Healthcare Technology Report's Top 25 Leaders in Medical Devices in 2021, and *NJBIZ*'s 2023 People to Watch in Finance lists. Purdue honored her with a Distinguished Engineering Alumna Award in 2024.

PRESIDENT OF THE SEATTLE MARINERS

Charles G. "Chuck" Armstrong earned his bachelor's degree in industrial engineering in 1964. He also received the G.A. Ross Award for Outstanding Senior Man. After earning his law degree from Stanford University, he spent over three years on active duty in the Navy, the highlight of which was qualifying as officer of the deck for flight operations on an aircraft carrier. Following naval service, Armstrong joined a Los Angeles-based law firm, making partner before taking a leave of absence to become president of a national furniture manufacturer and importer.

Armstrong later became chief executive officer of a real estate/development firm owned by the Seattle Mariners baseball team owner. In 1983, Armstrong was appointed president of the Mariners, a position he held until the team was sold in 1989. In 1991, he served as interim athletic director at the University of Washington, after which, at the request of a new Mariners ownership group, he returned to serve as president from 1992 to 2014. Armstrong played a significant role in turning around a struggling franchise, building a state-of-the-art retractable roof ballpark and leading the team to achieve a Major League Baseball regular season record of 116 wins in 2001.

FIRST TO WALK ON THE MOON

Neil Armstrong earned a bachelor's degree in aeronautical engineering in 1955 before joining NASA as an experimental research pilot. He became an astronaut in 1962 and served as command pilot for Gemini 8 in 1966, carrying a ceremonial Purdue flag into space. On July 20, 1969, Armstrong made history by becoming the first person to set foot on the moon, uttering the iconic words, "That's one small step for [a] man, one giant leap for mankind." Armstrong received an honorary doctorate from Purdue in 1970. Today at Purdue, the Neil Armstrong Hall of Engineering is named for him, adorned with a statue at its entrance and preserving his footprints from that momentous first moonwalk.

CREATOR OF THE MOST FREQUENTLY MANUFACTURED DEVICE IN HISTORY

"John" Mohammed M. Atalla earned a master's degree and a PhD in mechanical engineering in 1947 and 1949, respectively, establishing a solid foundation for his groundbreaking contributions to the semiconductor industry.

At AT&T Bell Laboratories, Atalla focused on enhancing the reliability of electromechanical relays. In 1959, he achieved a monumental breakthrough by co-inventing the MOSFET (metal-oxide-semiconductor field-effect transistor), a pivotal advancement in electronics and the most frequently manufactured device in history.

Three years later, Atalla co-founded Hewlett-Packard Associates, providing HP with crucial solid-state capabilities. In 1972, he founded Atalla Corporation, where he developed the Atalla Box—a revolutionary security system designed to encrypt PIN and ATM messages, integral to securing global ATM transactions. Renowned as one of the "Sultans of Silicon," Atalla significantly shaped the semiconductor industry. His contributions were recognized with a Benjamin Franklin Medal in Physics in 1975 and induction into the National Inventors Hall of Fame in 2009. Purdue awarded Atalla an honorary degree in 2003 and named the Atalla Institute for Advanced System Integration and Packaging for him.

LIBRARY OF CONGRESS

RECIPIENT OF THE DISTINGUISHED FLYING CROSS FOR HEROISM IN WORLD WAR II

John F. "Jack" Babbitt Jr.'s life journey is one of valor and engineering prowess. Drafted into the military during his first year at Purdue in 1942, he quickly rose to prominence, piloting a B-17 bomber in perilous missions over Nazi Germany as the lead plane in a squadron that numbered as many as 1,000 U.S. planes.

Despite overwhelming odds, Babbitt completed 30 combat missions, his aircraft bearing the scars of more than 220 bullet holes from enemy fire. After the war, he returned to Purdue, graduating with a bachelor's degree in chemical engineering in 1948.

He served as president of First Mississippi Corporation and Agrico Chemical Company. Babbitt founded Devco International and Devco Overseas Company to design, build and operate plants using the Kellogg Process to synthesize commercial chemicals.

The plants he constructed around the world have prevented the starvation of millions and made food more abundant by providing cheap and plentiful fertilizer, sparing more environmentally damaging mining methods. Purdue honored Babbitt with an Outstanding Chemical Engineer Award in 2020. The U.S. Armed Forces recognized his heroism with a Distinguished Flying Cross.

XEROX RECORD HOLDER FOR U.S. PATENTS

Santokh Badesha, a distinguished professor in the Elmore Family School of Electrical and Computer Engineering, has played a pivotal role in driving the digital publishing revolution. His inventions have profoundly impacted the productivity and communication of millions of individuals and businesses worldwide.

Since the 1990s, nearly all Xerox printing machines have incorporated his innovations. Badesha's inventions have been instrumental in popularizing laser printing technology and pioneering color laser printing. With an impressive portfolio of 273 U.S.-issued patents, he is one of the most prolific inventors in history.

He is a member of the National Academy of Engineering, a Fellow of the National Academy of Inventors, a Foreign Fellow of the Indian National Science Academy, a Fellow of the Royal Society of Chemistry and an Honorary Member of the Society for Imaging Science & Technology.

The Institution of Engineering and Technology honored him with the IET Mountbatten Medal in 2022. Badesha continues to fuel advances in digital printing, cementing his legacy as a visionary engineering inventor and leader.

INVENTOR OF THE BANBURY MIXER

Fernley H. Banbury graduated with a bachelor's degree in electrical engineering in 1906. Throughout his career, he worked exclusively for Farrel Corporation.

In 1916, Banbury received a U.S. patent for a groundbreaking mixer capable of blending a wide range of compounds. This invention, known as the Banbury Mixer, underwent testing by the Goodyear Company and marked a turning point in the company's evolution.

More than a century later, the Banbury Mixer remains in widespread use. Underscoring the invention's enduring impact, the Rubber Division of the American Chemical Society annually presents the Fernley H. Banbury Award to scientists and engineers who develop production equipment, control systems and instrumentation that significantly contribute to the rubber industry.

COURTESY OF SCIENCE HISTORY INSTITUTE

VIRGIN GALACTIC'S VP OF GOVERNMENT AFFAIRS AND RESEARCH OPERATIONS

Sirisha Bandla graduated with a bachelor's degree in aeronautical and astronautical engineering in 2011, laying the foundation for her extraordinary journey in the aerospace industry. Following graduation, she began her career at L-3 Communications as an aerospace engineer, playing a key role in designing components for advanced aircraft. Later, she moved to the Commercial Spaceflight Federation, dedicating her efforts to advancing the prospects of commercial flight.

In 2015, Bandla joined Virgin Galactic as vice president for government affairs, responsible for establishing Virgin Galactic's and Virgin Orbit's presence within the Beltway. She engaged extensively with Congress, the White House and NASA to promote the company's interests and objectives. Bandla's commitment and contributions to the aerospace sector culminated in a historic event on July 11, 2021, when she flew on the Virgin Galactic Unity 22 test flight alongside Sir Richard Branson and four others. This milestone made her the second Indian-born woman to fly to space. Bandla's achievements garnered widespread recognition. She was honored as one of the British Broadcasting Corporation's list of 100 inspiring and influential women around the world for 2022.

U.S. COMMISSIONER OF PATENTS AND TRADEMARKS

Donald W. Banner earned a bachelor's degree in electrical engineering in 1948. He pursued his education with the assistance of the GI Bill. At the young age of 19, Banner served as a fighter pilot and was shot down over Europe, enduring the hardships of being a prisoner of war until the end of World War II in 1945.

After completing his undergraduate studies at Purdue, Banner earned his JD from the University of Detroit, followed by a Master of Patent Law and a Doctorate of Laws from the John Marshall Law School.

A successful patent attorney, Banner was admitted to practice before the Supreme Court and served for 15 years as BorgWarner's general patent counsel. In 1972, he was appointed U.S. commissioner of patents and trademarks, and chairman of the American Bar Association's Patent, Trademark and Copyright Law section. He received a Jefferson Medal in 1984 for his contributions to patent, trademark and copyright laws, and Purdue presented him with an Outstanding Electrical Engineer Award in 1994.

PIONEER IN DIGITAL CAMERA IMAGE PROCESSING

Farhan A. Baqai earned a master's degree and a PhD in electrical engineering in 1997 and 2000, respectively. He currently serves as a senior research manager at Apple. Prior to his role at Apple, Baqai held research and leadership positions at the Sony U.S. Research Center and Xerox Corporation.

He and his teams have made significant contributions to key technologies for the iPhone, the iPad, the Mac and Sony's CyberShot and Alpha digital cameras. His extensive expertise spans various fields, including digital camera processing, machine learning, computer vision, stereoscopic image processing, statistical signal processing, digital printing and radar imaging. In recognition of his contributions, the Institute of Electrical and Electronics Engineers honored Baqai as a fellow in 2023. Additionally, Purdue presented him with an Outstanding Electrical and Computer Engineer Award in 2021.

SAS SCANDANAVIAN AIRLINES

The prototype 747 was first displayed to the public on Sept. 30, 1968, at the Paris Air Show.

LEADER OF BOEING 747 DEVELOPMENT

Robert E. "Bob" Bateman, who earned a bachelor's degree in aeronautical engineering in 1946 and an honorary doctorate in 1992, dedicated his entire 42-year career to The Boeing Company, leaving an enduring mark on the aerospace industry.

He began by contributing to development of the Boeing B-52 and 707 and later led pioneering projects in reentry vehicles, advanced marine vehicles and lunar exploration. Bateman's most renowned achievement was leading the billion-dollar development of the Boeing 747, an aviation milestone.

He held various leadership roles at Boeing, including vice president of the company's Washington, D.C., office, as well as vice president of governmental and institutional affairs. Purdue recognized Bateman with a Distinguished Engineering Alumnus Award in 1974, an Outstanding Aerospace Engineer Award in 1999, a Purdue Engineering Alumni Association Presidents Lifetime Award in 1998, and the Order of the Griffin in 2005. Additionally, Bateman was inducted into the Navy League Hall of Fame, and the secretary of the navy presented him with a Meritorious Public Service Award in 1968 and a Distinguished Public Service Award in 1972.

VICE COMMANDER OF THE U.S. NAVAL FORCES CENTRAL COMMAND

Retired Rear Adm. **Douglas "Woody" Beal** earned a bachelor's degree in aeronautics and astronautics engineering in 1986. Following graduation, he entered the Aviation Officer Candidate School and earned his gold wings in 1988. Beal subsequently served in the U.S. Navy as a fighter pilot, accumulating 3,844 military flight hours, participating in 55 combat sorties and completing 472 carrier-arrested landings.

In 2017, Beal was appointed vice commander of the U.S. Naval Forces Central Command/U.S. Fifth Fleet. Additionally, he has worked as a pilot for American Airlines, amassing more than 9,500 mishap-free miles. Purdue presented Beal with an Outstanding Aerospace Engineer Award in 2019.

The Bechtel Innovation Design Center opened in 2017. Students may use the Bechtel Center to work on class assignments, develop club projects, build design competition entries, create a product prototype for an entrepreneurial endeavor, or design innovative solutions for their own enjoyment.

RECIPIENT OF A NATIONAL MEDAL OF TECHNOLOGY AND INNOVATION

Stephen Bechtel Jr. graduated with a bachelor's degree in civil engineering in 1946 and joined his family's business, Bechtel Corporation, in 1948. Within 12 years, he gained experience on projects across the U.S. and Canada. He became chief executive officer in 1960 and chairman in 1969, overseeing Bechtel's significant growth over three decades and leading a workforce of 44,000 employees. During his tenure, Bechtel secured major civil engineering projects worldwide, including the Channel Tunnel and King Khalid International Airport and Jubail Industrial City, both in Saudi Arabia. From 1982 to 1986, Bechtel served as chairman of the National Academy of Engineering. He received numerous accolades for his contributions to engineering, such as a Hoover Medal in 1980, elevation to National Honor Member of Chi Epsilon in 1990 and an Honor Award from the National Building Museum in 1998. In addition, Bechtel earned a National Medal of Technology and Innovation in 1991, and Purdue's Bechtel Innovation Design Center bears his name.

DESIGNER OF WORLD WAR II FIGHTER PLANES

Donovan R. Berlin earned his bachelor's degree in mechanical engineering in 1921 and was awarded an honorary doctorate in 1953. He began his aviation career at Douglas Aircraft Company in 1926 and joined Northrop Corporation in 1929 before moving to Curtiss-Wright in 1934.

At Curtiss-Wright, Berlin became renowned for his contributions to military aviation worldwide, designing fighter planes noted for their safety, durability and pilot-friendliness. Among his design accomplishments, the P36 and P40 fighter planes became two of the most widely used aircraft in World War II.

NATIONAL ARCHIVES

DEPUTY DIRECTOR OF DARPA'S TACTICAL TECHNOLOGY OFFICE

Thomas Beutner received a bachelor's degree in aeronautical and astronautical engineering in 1987. After pursuing advanced degrees at Stanford University, he launched a career that propelled innovation in the U.S. Department of Defense, resulting in enhanced aircraft and weapons systems.

In 1993, Beutner joined the Air Force Research Laboratory, focusing on advanced diagnostics. Two years later, he became program manager for turbulence and rotating flows in the Air Force Office of Scientific Research. In 2005, Beutner transitioned to the Defense Advanced Research Projects Agency (DARPA) as project manager in its Tactical Technology Office.

He furthered his career by joining the Office of Naval Research in 2010 as director of the Aerospace Sciences Research Division and becoming department head of the Air Warfare and Weapons Department in 2014. Returning to DARPA in 2018 as deputy director of the Tactical Technology Office, Beutner contributed substantially to the DOD.

In 2022, he joined the faculty at the University of Maryland as a research professor. Purdue honored Beutner with a Distinguished Engineering Alumnus Award in 2018 and inducted him into the Cooperative Education Hall of Fame.

INVENTOR OF JUNE OVEN AND MANY APPLE INNOVATIONS

Nikhil Bhogal graduated with a bachelor's degree in computer engineering in 2003. He started his career at Motorola, where he worked on the camera system of the iconic Moto RAZR. Bhogal then joined Apple as one of its first camera engineers, contributing to the development of the original iPhone through the iPhone 5.

At Apple, he is credited with numerous imaging technology patents, including innovations such as tap-to-focus, panorama processing, continuous image capture, zero shutter lag and the lock screen camera.

To expand the use of camera technology beyond photography, Bhogal co-founded June, a company combining deep learning and computer vision to revolutionize cooking. The June Intelligent Oven can autonomously recognize and cook food while integrating the functionality of 12 appliances into one.

In 2021, June was acquired by Weber Grills, where Bhogal currently serves as senior vice president of technology and connected devices while continuing as June's co-founder and chief technology officer.

Birck Nanotechnology Center

FOUNDER AND CEO OF TELLABS

Michael J. Birck, a 1960 graduate with a bachelor's degree in electrical engineering, had a remarkable career. He commenced his professional journey by joining the AT&T Bell Laboratories technical staff and then worked for Continental Telephone Laboratories and Wecom Inc. before venturing into entrepreneurship.

In 1974, Birck founded Tellabs, a company initially focused on manufacturing echo suppressors to eliminate annoying echoes on phone calls. Tellabs experienced significant growth and success, going public in 1980 and expanding to employ more than 2,000 people across 25 countries by 1990. With Birck as president and CEO until his retirement in 2004, Tellabs emerged as a global leader in producing voice and data networking products.

Birck received numerous accolades, including the IEEE Ernst Weber Engineering Leadership Award in 2001, the Electronics Industry Alliance's Medal of Honor Award in 2003 and Purdue's Distinguished Engineering Alumnus Award in 1991. His legacy lives on at Purdue through the Birck Nanotechnology Center and the Birck Boilermaker Golf Complex.

INSPIRATIONAL TRAILBLAZER AND MENTOR FOR MINORITY ENGINEERS FOR MORE THAN 30 YEARS

Marion Williamson Blalock, an impactful pioneer, advocate, mentor and advisor, directed and served the Minority Engineering Program (MEP) from its inception in 1974 to 2008. She created and led recruitment, retention and graduation initiatives that helped more than 2,300 underrepresented minority students earn engineering degrees during her tenure and paved the way for later students. Numerous Purdue MEP students have gone on to stellar careers in engineering, education and business. In addition, MEP has become a model for universities across the nation. A Purdue alumna (bachelor's degree in sociology and psychology in 1969 and master's degree in counseling and personnel services in 1973), Blalock also served as a National Society of Black Engineers (NSBE) advisor and board member.

PILOT ON THE FIRST FLIGHT OF AMERICA'S B-1 BOMBER

Col. **Charles C. "Charlie" Bock Jr.** earned a bachelor's degree in aeronautical engineering in 1949. He had a long and distinguished career as a bomber pilot, fighter pilot, test pilot and military astronaut designee.

Bock flew more than 100 combat missions during the Korean and Vietnam wars. In 1973, he retired from the U.S. Air Force and joined Rockwell International Corporation as chief test pilot for the B-1 bomber program. He piloted the first test flight of the B-1 bomber in December 1974.

After retiring from Rockwell in 1981, Bock served Northrop Corporation from 1982 to 1984 as a consultant on its B-2 stealth bomber project. In recognition of his accomplishments, Purdue honored Bock with an Outstanding Aerospace Engineer Award in 2016.

FOUNDER OF COMPANY THAT HAS BUILT 700-PLUS WATER TREATMENT PLANTS IN AMERICA

Robert L. "Bob" Bowen earned a bachelor's degree in civil engineering in 1962 and an honorary doctorate in 2007. His remarkable career has significantly advanced the construction industry and his alma mater.

After graduating, Bowen honed his skills as a project manager at W.M. Lyles Co. in Fresno, California, until 1967, when he returned to Indiana to establish the Bowen Engineering Company. For the next 42 years, Bowen served as CEO of his company, specializing in both traditional and engineering procurement and construction (EPC) projects in the water/wastewater, energy and industrial sectors. He continues as chairman.

Employing more than 300 professionals, Bowen Engineering has built 700-plus water treatment plants in the U.S. The company maintains a strong connection with Purdue, consistently hiring its engineers and supporting the university's initiatives. Bowen's dedication to Purdue has extended beyond his professional endeavors, encompassing organizing an annual bike ride from Indianapolis to West Lafayette for Boilermaker home football games for more than two decades, as well as teaching classes as the Donn E. Hancher Distinguished Fellow in the Lyles School of Civil and Construction Engineering.

Bowen has received numerous accolades, including the American Society of Civil Engineers' Mark Benjamin Lifetime Achievement Award in 2018, induction into the Indiana Public Schools Hall of Fame in 2009 and the Indiana Heartland Entrepreneur of the Year Award from Ernst & Young in 1998. Purdue presented Bowen with an Outstanding Civil Engineer Award in 1995 and a Distinguished Engineering Alumnus Award in 1999.

Today, Bowen's name remains prominent at Purdue through the Robert L. and Terry L. Bowen Laboratory for Large-Scale Civil Engineering Research and scholarships awarded to students in his honor.

DIRECTOR OF NASA'S MARSHALL SPACE FLIGHT CENTER

G. Porter Bridwell earned a bachelor's degree in aeronautical and astronautical engineering in 1958. His engineering career began at Rocketdyne, where he worked on rocket engine design. In 1962, he commenced a 34-year tenure at NASA's Marshall Space Flight Center.

Bridwell initially worked in the Saturn Systems Office and the Saturn V Program. In 1975, he transitioned to the Shuttle Systems Office, holding such roles as chief of the Project Engineering Office and deputy manager of the external tank project. Bridwell became manager of the external tank project in 1983 and advanced to manager of the Shuttle Projects Office in 1989. A year later, he was appointed director of national launch systems at NASA headquarters.

In 1992, Bridwell returned to the Marshall Center as manager of the Heavy Lift Launch Vehicles Office, overseeing design, development and integration efforts. During his career, he also undertook special assignments with the Space Shuttle redesign team and, later, the U.S./Russia Space Station integration team.

In 1994, Bridwell became the seventh director of NASA's Marshall Space Flight Center, serving until his retirement in 1996. Purdue recognized Bridwell with an Outstanding Aerospace Engineer Award in 2013.

ASTRONAUT WHO ORBITED THE EARTH 161 TIMES ON SPACE SHUTTLE FLIGHTS

Mark N. Brown graduated with a bachelor's degree in aeronautical and astronautical engineering in 1973. He then joined the U.S. Air Force, earning his pilot wings a year later. Over the next six years, Brown flew T-33 and F-106 aircraft, obtained a master's degree from the Air Force Institute of Technology and completed the Air Command and Staff College. He moved to NASA's Johnson Space Center to work on Space Shuttle contingency procedures in 1980 and became an astronaut in 1984.

In the latter role, Brown orbited the Earth 161 times and logged more than 10 days in space, spanning two missions: Columbia (STS-28) in 1989 and Discovery (STS-48) in 1991. After retiring from NASA in 1993, he held senior leadership positions at General Research Corporation, AT&T Government Solutions, Computer Sciences Corporation, MTC Technologies, BAE Systems and Peerless Technologies Corporation. Brown, a retired Air Force colonel, received numerous honors during his career, including the Air Force Command Pilot designation and NASA Space Flight Medal.

LEADER OF BOEING'S 8,000-PLUS ENGINEERS

Mark A. Burgess earned three Purdue degrees: bachelor's (1978) and master's (1979) degrees in aeronautical and astronautical engineering, as well as a master's degree in industrial administration (1982).

He began his career at The Boeing Company in 1979, spending his first 14 years in various engineering roles. In 1993, as part of the Joint Strike Force, he became site director for St. Louis and team leader on the X-32 air vehicle analysis and integration.

Burgess was elevated to director of Advanced Lean Enterprise and chief information officer for Phantom Works eight years later and promoted to chief engineer for Phantom Works in 2007. Two years after that, he became chief engineer for Boeing Research and Technology. In 2010, he assumed the chief engineer role for engineering operations and technology, taking charge of Boeing's centrally managed engineering and development operations and executive management of more than 8,000 engineers. Purdue honored Burgess with a Distinguished Engineering Alumnus Award in 2013.

ONE OF THE UNITED STATES' TOP EXPERTS IN STORMWATER MANAGEMENT

Christopher B. Burke earned three civil engineering degrees from Purdue: a bachelor's degree (1977), a master's degree (1979), and a PhD (1983). In 1986, he founded Christopher B. Burke Engineering, Ltd. (CBBEL), which has grown into one of the top consulting engineering firms in the region, boasting nearly 250 employees.

While Burke is renowned for his expertise in stormwater management and water resources, his firm offers a wide range of engineering, surveying and environmental services to clients in the public and private sectors. Burke has received numerous honors, including membership in the National Academy of Engineering and National Academy of Construction and being named a distinguished member of the American Society of Civil Engineers.

Purdue presented Burke with a Distinguished Engineering Alumnus Award in 1998 and an honorary doctorate in 2010. He also received an honorary doctorate from the University of Illinois in 2021.

At Purdue, his generosity is reflected in the Christopher and Susan Burke Hydraulics and Hydrology Research and Teaching Laboratory, Christopher B. and Susan S. Burke Distinguished Professor of Civil Engineering, and the Christopher B. and Susan S. Burke Graduate Program in Civil Engineering.

Christopher and Susan Burke Hydraulics and Hydrology Research and Teaching Laboratory

CELEBRATING OUR CONSEQUENTIAL IMPACT ON THE WORLD

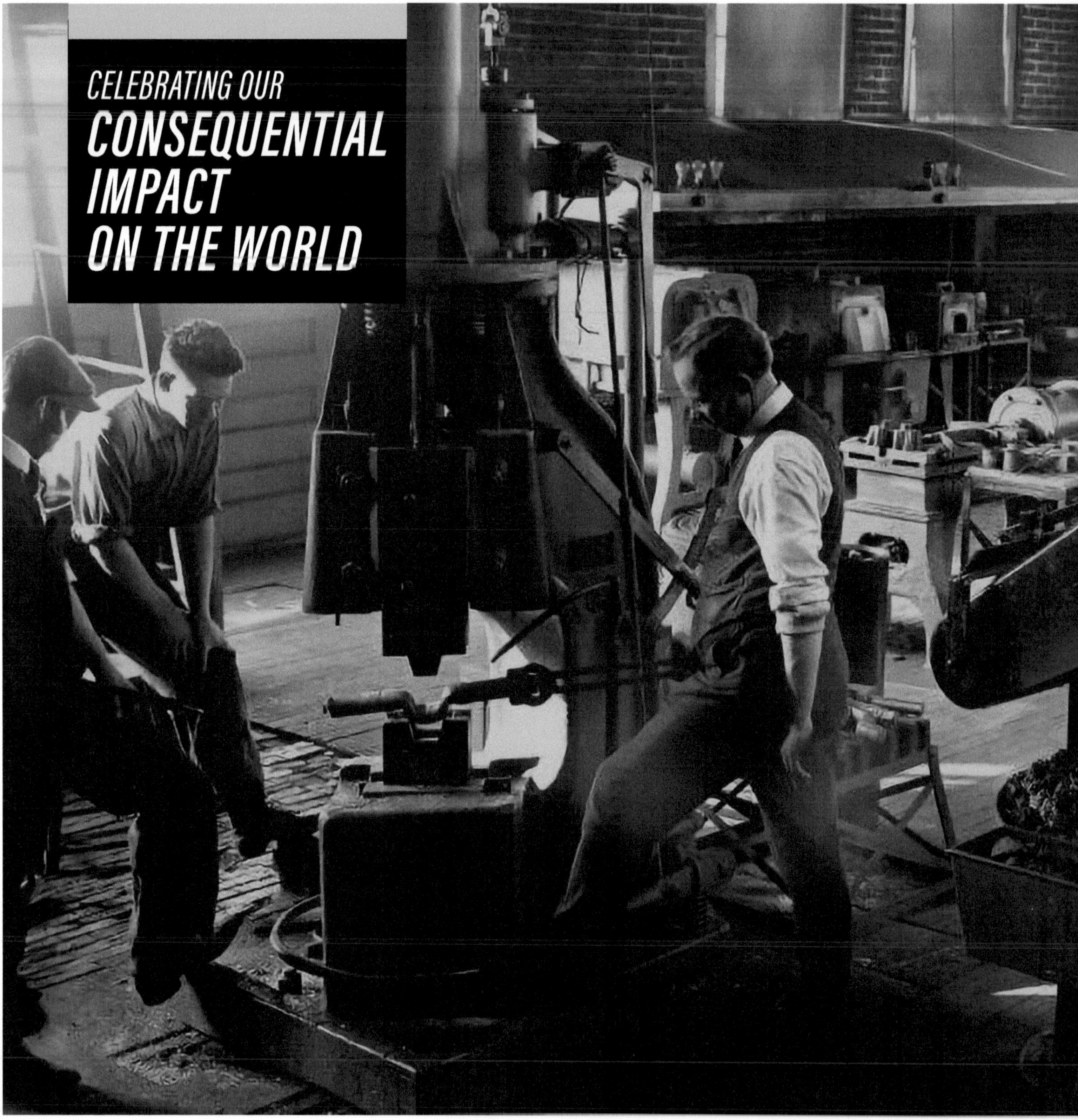

1920s

Students work in the forge located in the Michael Golden Laboratories and Shops. Constructed in 1910 as the Practical Mechanics Building, the facility was renamed 11 years later "to commemorate the life of one who from 1884 to 1917 devoted himself untiringly to the university and its students."

CREATOR OF WEARABLE SENSING TECHNOLOGY

Stacey Burr earned a bachelor's degree in industrial engineering in 1984 and a master's degree in industrial administration from Purdue in 1991. The West Lafayette native has built a very successful career in wearable sensing technology.

After beginning her journey at DuPont, Burr founded and served as CEO of Textronics Inc., where she led commercialization of e-textiles, creating the first wearable sensor garments for fitness and health. Burr later sold Textronics to Adidas and continued as vice president of wearable sports electronics.

A few years later, Google recruited her to take on a new role—as vice president of product & UX for Wear OS and Google Fit. Most recently, she started her own company, Future Standard Inc., a consulting and advising firm in the health tech sector. During her career, Burr has co-invented 76 patented materials and process technology innovations in wearable electronics. Purdue honored her as an Old Master in 2020.

ORIGINATOR OF THE WORD "ELECTRONICS"

Orestes Hampton Caldwell graduated with a bachelor's degree in electrical engineering in 1908. He commenced his career as the technical editor of *Western Electrician,* and later edited such prominent engineering and technical journals as *Electrical World, Electrical Merchandising* and *Radio Retailing.*

In 1927, with the endorsement of then-U.S. Department of Commerce Secretary Herbert Hoover, Caldwell was appointed to the newly established Federal Radio Commission, where he garnered a reputation as the most technologically proficient commissioner. Known for his passion for electronic gadgets, Caldwell is credited with coining the term "electronics."

MOST RECENT TO WALK ON THE MOON

Eugene A. "Gene" Cernan earned a bachelor's degree in electrical engineering in 1956 and received an honorary doctorate in 1970. Following graduation, he embarked on a remarkable career in the U.S. Navy, where he attended flight school and became a naval aviator.

Cernan amassed an impressive flying record, logging more than 5,000 hours of flight time and completing at least 200 successful aircraft carrier landings. In 1963, NASA selected him for the third group of astronauts for the Gemini and Apollo space programs. During the Gemini 9 mission, Cernan made history as the second American to walk in space.

He piloted Apollo 10, the dress rehearsal for the historic Apollo 11 mission, bringing the spacecraft within a mere 8.5 miles of the moon's surface. Notably, Apollo 10 set the record for the highest speed achieved by any crewed vehicle during its return to Earth—a staggering 24,791 mph. In a crowning achievement, as commander of Apollo 17 in 1972, Cernan became the most recent person to set foot on the moon. He etched his daughter's initials, TDC, onto the lunar surface before departing. Cernan set the lunar surface speed record of 11.2 mph in the Lunar Roving Vehicle.

His life story has been immortalized in books and movies, cementing his place as an American legend. Purdue recognized Cernan with an Outstanding Electrical and Computer Engineer Award in 1992.

Apollo 17 mission commander Eugene Cernan drives the Lunar Roving Vehicle during the early part of the mission's first moonwalk at the Taurus-Littrow landing site. The Lunar Module is in the background.

"THE SHOT PEENING MAESTRO"

Jack Champaigne obtained a bachelor's degree in electrical engineering in 1968. After gaining experience in related fields, he founded Electronics Inc. in 1974. The company swiftly rose to prominence, becoming a global authority on shot peening, a surface enhancement process, for the aerospace, automotive, medical and energy industries.

Champaigne's pioneering work led to the development of the MagnaValve, a groundbreaking magnetic media valve, and the acquisition of 30 U.S. patents related to shot peening technology. In addition to marketing products used worldwide, Champaigne has been committed to education and the sharing of information. In 1991, he spearheaded the inaugural Shot Peening and Blast Cleaning Workshop, a milestone training event in the industry.

He was inducted into the International Scientific Committee on Shot Peening in 1993 and started publication of *The Shot Peener Magazine* in 1986. Champaigne also played a vital role as a member and past chairman of the International Scientific Committee on Shot Peening and the Society of Automotive Engineers. He was selected as the 2024 Shot Peener of the Year at the annual Shot Peening Workshop in Colorado Springs, Colorado. Champaigne's experience and thorough understanding of the history, theory, specifications and practical applications of shot peening provide an invaluable contribution to the industry, earning him the title, "The Shot Peening Maestro."

THE 13TH PRESIDENT OF PURDUE UNIVERSITY

Mung Chiang is the 13th president of Purdue University and the Roscoe H. George Distinguished Professor of Electrical and Computer Engineering. He was previously the John A. Edwardson Dean of the College of Engineering and executive vice president for strategic initiatives at Purdue University, as well as the Arthur LeGrand Doty Professor of Electrical Engineering at Princeton University.

Chiang founded the Princeton EDGE Lab in 2009 and then founded several startup companies and industry consortia in edge computing. As the science and technology advisor to the U.S. secretary of state in 2020, he initiated tech diplomacy programs for the U.S. government. Currently, he serves on the inaugural board of the U.S. Foundation for Energy Security and Innovation and several corporate and nonprofit boards.

Chiang earned his BS (1999), MS (2000) and PhD (2003) from Stanford University and received an honorary doctorate (2024) from Dartmouth College. For his research in communication networks, he received the Alan T. Waterman Award (2013) from the National Science Foundation; the Kiyo Tomiyasu Award (2012) and the INFOCOM Achievement Award (2022) from the Institute of Electrical and Electronics Engineers; and a Guggenheim Fellowship (2014). He was elected to the American Academy of Arts and Sciences (Class of Mathematical and Physical Sciences 2024), the National Academy of Inventors (2020), and the Royal Swedish Academy of Engineering Sciences (2020).

CHAIRMAN AND CEO OF CATALENT —A COMPANY WITH 18,000 EMPLOYEES

John R. Chiminski obtained a master's degree in electrical engineering in 1987. His corporate career has had an enduring impact on the health care and pharmaceutical industries. Upon graduating, Chiminski joined GE HealthCare, where he spent 21 years ascending the corporate ladder to president and chief executive officer of GE Medical Diagnostics in 2007.

His prior roles included vice president and general manager of GE Healthcare Services, as well as vice president and general manager of GE Magnetic and Resonance Business. Chiminski's exceptional management skills led him to Catalent Pharma Solutions, where he was named president and CEO in 2009 and appointed chairman of the board in 2016, serving until his retirement in 2022.

His tenure at Catalent saw significant growth and innovation, solidifying the company's position as a trailblazer in the pharmaceutical industry. Catalent today has about 18,000 employees. Currently, Chiminski serves on the board of directors of PhenomeneX, contributing his expertise to further advancements in health care.

Purdue recognized Chiminski with an Outstanding Electrical and Computer Engineer Award in 2006 and a Distinguished Engineering Alumnus Award in 2012. A generous gift from Chiminski and his wife, Laura A. Chiminski, enabled the College of Engineering to establish the Chiminski Collaborative Research Hub.

THE BRAINS BEHIND THE MEMRISTOR

Leon O. Chua was a Purdue electrical engineering faculty member from 1964 to 1970. He is the inventor and namesake of Chua's circuit, one of the first and most widely known circuits to exhibit chaotic behavior. Chua also was the first to conceive the theories behind, and postulate the existence of, the memristor.

Thirty-seven years after Chua predicted its existence, a working solid-state memristor was created by a team at Hewlett-Packard. Memristors can recognize and "remember" the amount of electrical current that passes through them and have the potential to revolutionize integrated circuit technology, possibly even leading to processors that can mimic the neural and synaptic functions of the human brain.

Today, the memristor is used for many applications, including digital memory, neuromorphic systems and logic circuits. The memristor has the potential to provide low-energy consumption artificial intelligence. It has become the fourth fundamental electric circuit element, offering applications throughout electrical engineering.

A FOUNDER OF THE SOCIETY OF BLACK ENGINEERS AT PURDUE

Fred Cooper grew up in the Chicago area and attended Purdue University between 1970 and December 1974 on a full football scholarship. He was a three-year starter for the Boilermakers and was elected defensive captain his senior year.

He graduated from Purdue in December 1974 with a bachelor of science in Electrical Engineering. In 1971, during his freshman year, Cooper and his classmate Edward Barnette (now deceased) approached the dean of engineering with the idea of creating the Society of Black Engineers (SBE). Their goal was to improve the recruitment and retention of Black engineering students at Purdue.

The organization grew from about 15 students and later became the precursor to the National Society of Black Engineers (NSBE) in 1975. Cooper served as president of SBE during the 1972–73 school year.

Today, NSBE is one of the largest student-governed organizations in the U.S., with more than 31,000 members. After graduating from Purdue, Cooper was drafted by the Detroit Lions and played as a defensive back for two seasons.

After his football career, he joined the Bell System, where he spent 25 years in engineering, product marketing, project management, manufacturing planning, technical sales support, international market development and product management.

Cooper has continued to serve in leadership roles within the Purdue Engineering Alumni Association and the Minority Engineering Program. He served on the Purdue Engineering Alumni Association Board of Directors from 2017–23 and was elected president of the organization in 2021–22. Cooper's legacy at Purdue has inspired and served as a role model for engineering students across the country.

PRINCIPAL DESIGNER OF AT&T'S FIRST CMOS MICROPROCESSOR

James A. "Jim" Cooper Jr. earned a PhD in electrical engineering in 1973. Afterward, he joined the technical staff at AT&T Bell Laboratories, where he worked for the next 10 years. At Bell Labs, Cooper designed AT&T's first CMOS microprocessor and developed a time-of-flight technique for investigating high-field transport in silicon inversion layers.

In 1983, he joined engineering faculty at Purdue, where he became founding director of the Purdue Optoelectronics Research Center and the Jai N. Gupta Professor of Electrical and Computer Engineering, and he played a key role in development and direction of Purdue's Birck Nanotechnology Center.

At Purdue, Cooper has focused on device technology in the wide bandgap semiconductor SiC. His group demonstrated several significant advancements, including the first monolithic integrated circuits in SiC (1993), the first planar DMOS power transistors (1996), the first lateral DMOSFETs (1997) and the first self-aligned short-channel DMOSFETs (2003). Cooper holds 18 U.S. patents.

John Costas (at right) discusses the Synchronous Receiver A/N FRR-48 XW-1 with GE employee Bill Jones.

INVENTOR OF THE COSTAS LOOP AND COSTAS ARRAY

John Peter Costas earned bachelor's and master's degrees in electrical engineering in 1944 and 1947, respectively. After obtaining a PhD from the Massachusetts Institute of Technology in 1951, Costas joined General Electric as an engineer. At GE, he developed a modified phased-locked loop that recovers the "suppressed" carrier in many digital communications receivers, known as the Costas loop.

This invention had a profound effect on wireless receivers and modern digital communications. In the 1960s, Costas played a pivotal role in solving the mystery concerning the poor performance of sonar systems. He discovered that the rapidly time-varying channel made coherent processing inappropriate. His solution involved using a permutation array, now known as a Costas array, which possesses ideal properties to address the problem.

DEVELOPMENT LEAD FOR NASA'S HYPERSONIC AND TRANSATMOSPHERIC FLIGHT

Lana M. Couch graduated with a bachelor's degree in aeronautical engineering in 1963. She began her career as a research engineer at NASA's Langley Research Center.

Over the next 30 years, she alternated between assignments at Langley and NASA headquarters in Washington, D.C., steadily advancing through the management ranks. Her accomplishments included leading development of NASA's hypersonic and transatmospheric flight.

Couch's career culminated in her appointment as director of the National Aero-Space Plane Office at Langley. She received numerous accolades, such as a Women in Aerospace Lifetime Achievement Award, a NASA Exceptional Achievement Medal and a National Aviation Club's Women in Aviation Award. Purdue recognized Couch with a Distinguished Engineering Alumna Award in 1994.

CREATOR OF COLLABORATIVE SOFTWARE KNOWN AS THE WIKI

Ward Cunningham graduated in 1978 with a bachelor's degree in interdisciplinary engineering, focused on electrical engineering and computer science. He is recognized for creating collaborative software that evolved into what is now known as the wiki. In 1995, Cunningham launched the WikiWikiWeb application, which gained significant popularity, especially through its most famous implementation, Wikipedia, with hundreds of millions of users worldwide. Additionally, Cunningham is credited with "Cunningham's Law," which posits that "the best way to get a right answer on the internet is not to ask a question, but to post the wrong answer." This observation highlights his belief that people are more likely to correct misinformation than to answer a question directly.

NASA'S CHIEF TECHNOLOGIST–OVERSEEING $1B-PLUS IN INVESTMENTS

John Dankanich earned bachelor's and master's degrees in aeronautics and astronautics in 2001 and 2003, respectively. His 20-year career at NASA has spanned various roles, including design, build, test, system analysis, technology maturation and project management.

Currently, Dankanich serves as chief technologist for NASA's Marshall Space Center, leading a workforce of more than 6,000 civil servants and overseeing a portfolio of investments exceeding $1 billion. These investments encompass a broad spectrum, ranging from basic research to flight system delivery.

Within NASA, Dankanich is esteemed as a subject matter expert in technology development, propulsion testing and qualification, low-thrust trajectory optimization, mission design and planetary defense. Purdue recognized him with an Outstanding Aerospace Engineer Award in 2022.

ONLY PURDUE FACULTY MEMBER ELECTED TO BOTH NAE AND NAS

Supriyo Datta, the Thomas Duncan Distinguished Professor of Electrical and Computer Engineering, is the only Purdue faculty member ever honored with membership in both the National Academy of Engineering and the National Academy of Sciences. His innovative proposals have launched new research areas like negative capacitance devices, molecular thermoelectricity and probabilistic bits.

Datta's groundbreaking work introduced spin-orbit coupling as a means to control electron spin using an electric field rather than a magnetic field. This contribution launched the field of spintronics, which is widely used in quantum computing.

Datta's research on current flow in nanoscale electronic devices led to the development of a novel approach to quantum transport modeling. By combining nonequilibrium Green's functions (NEGF) formalism from many-body physics with the Landauer formalism from mesoscopic physics, Datta pioneered a methodology that leading semiconductor technology companies adopted in computer-aided design tools. This approach is also used by quantum chemists in molecular electronics research.

Through his books and online courses, he has made his work accessible to a broad audience of engineers, physicists and chemists. In addition to NAE and NAS membership, Datta has received the William Procter Prize for Scientific Achievement from Sigma Xi and many other recognitions from the Institute of Electrical and Electronics Engineers, the American Physical Society, the Semiconductor Industry Association and the Semiconductor Research Corporation.

CEO OF BOEING'S PHANTOM WORKS

Darryl W. Davis graduated in 1978 with a bachelor's degree in aeronautical and astronautical engineering. His aerospace career was nothing short of remarkable. Joining McDonnell Douglas' propulsion department after graduation, Davis quickly distinguished himself with his expertise and innovative approach. His tenure spanned the McDonnell Douglas merger with The Boeing Company, where he continued to excel for 38 years.

In 2009, his exceptional leadership skills resulted in an appointment as CEO of Boeing's Phantom Works. In this role, Davis oversaw nearly 3,000 employees dedicated to developing cutting-edge products and technologies for the U.S. military, contributing notably to advancements in military aviation and defense capabilities.

Davis retired from Boeing in 2017, leaving a legacy of innovation and excellence in aerospace engineering. Purdue honored him with a Distinguished Engineering Alumnus Award in 2010.

CTO FOR HYUNDAI'S ADVANCED AIR MOBILITY VENTURE

Benjamin L. "Ben" Diachun earned a bachelor's degree in mechanical engineering in 1999. He furthered his education at Stanford University, earning a master's in aeronautics and astronautics. For the initial 16 years of his career, Diachun was associated with Scaled Composites, where he rose from engineer to president.

During his tenure, Diachun played a pivotal role in developing 15 distinct aerospace vehicle types. He also facilitated the sale of more than $1 billion in vehicle hardware and services while boosting annual revenue from $25 million to more than $135 million. In 2019, Diachun assumed the CEO role at Opener, a Silicon Valley startup focused on developing a personal all-electric air vehicle.

In two years, Hyundai recruited him to lead a similar initiative. Diachun thus became chief technology officer for Supernal, Hyundai's Advanced Air Mobility venture aimed at introducing new air vehicle product lines, including electric vertical take-off and landing (eVTOL) aircraft slated for market release in 2028. Purdue recognized Diachun with an Outstanding Mechanical Engineer Award in 2019 and a Distinguished Engineering Alumnus Award in 2023.

PURDUE UNIVERSITY ARCHIVES AND SPECIAL COLLECTIONS

In 1936, Amelia Earhart purchased this new Lockheed 10-E Electra with funds from the Purdue Research Foundation. The airplane was officially called the "Flying Laboratory" because it was meant to be used as a test bed for new equipment. But Earhart's real intent was to fly it around the world.

AMERICA'S "QUEEN OF THE AIR"

Amelia Earhart, a career counselor and advisor in Purdue's Department of Aeronautics from 1935 to 1937, was one of the most influential pioneers in American aviation. In 1932, she achieved the remarkable precedent of becoming the first woman to fly solo across the Atlantic Ocean, prompting her to be dubbed "Queen of the Air."

Earhart also found success as a bestselling author and a highly-sought-after product endorser. While at Purdue, she began planning the round-the-world flight that led to her legendary disappearance in 1937.

Today, Purdue has one of the most extensive collections of Earhart's papers, writings and artifacts, along with a residence hall named in her honor and a larger-than-life statue. In Spring 2024, ground was broken for the Amelia Earhart Terminal at the Purdue University Airport, which reopened to commercial flights in Summer 2024.

EXECUTIVE DIRECTOR OF THE AIR FORCE RESEARCH LABORATORY—WITH $4.4B BUDGET

C. Douglas Ebersole graduated with a bachelor's degree in aeronautics and astronautics in 1982. He had a distinguished 35-year career with the U.S. Air Force, holding key civilian leadership positions. Notably, he served as chief engineer and then director of engineering on the F-22 Program, director of engineering for the F-35 Joint Program Office and director of the Aerospace Systems Directorate, Air Force Research Laboratory.

Ebersole's final position was executive director of the Air Force Research Laboratory at Wright-Patterson Air Force Base, overseeing a workforce of more than 6,000 professionals and its $4.4 billion science and technology budget. In 2017, the president of the United States conferred on him the rank of Meritorious Executive in the Senior Executive Service for sustained superior performance.

Purdue honored him with an Outstanding Aerospace Engineer Award in 2014, a Distinguished Engineering Alumnus Award, and induction into its Cooperative Education Hall of Fame in 2016. In retirement, Ebersole continues to serve on corporate boards and is a member of the NASA Advisory Council, Human Exploration and Operations Committee.

TOP EXECUTIVE OF MANY GLOBAL CORPORATIONS

John A. Edwardson received a bachelor's degree in industrial engineering in 1971 and an honorary doctorate in 2006. After completing his studies, he launched a remarkable career leading major global corporations.

From 2001 to 2011, Edwardson was chairman and chief executive officer of CDW, a leading provider of technical products and services for business, government, and education. Under his leadership, CDW's annual revenue grew from $3.8 billion to more than $10 billion. Edwardson remained chairman until his retirement in 2012.

Earlier in his career, he was president of United Airlines, chief financial officer of Northwest Airlines, and chairman and chief executive officer of Burns International Services. Edwardson served on the board of directors of CDW, United Airlines, Burns International, Chubb Ltd, FedEx, and Telecom Corporation of New Zealand. Purdue honored Edwardson with an Outstanding Industrial Engineer Award in 1997 and a Distinguished Engineering Alumnus Award in 1988.

Charles Alton Ellis began working on the Golden Gate Bridge in 1922, the same year he gained his engineering certificate from University of Illinois.

ISTOCK/DAVID CHANG

STRUCTURAL DESIGN ENGINEER FOR THE GOLDEN GATE BRIDGE

Charles Alton Ellis, a civil engineering professor at Purdue from 1934 to 1946, contributed significantly to the design of San Francisco's Golden Gate Bridge.

The American Society of Civil Engineers cited the Golden Gate Bridge as one of the "Seven Wonders of the Modern World" in 1994, highlighting its engineering significance.

In 2012, as part of the bridge's 75th anniversary celebrations, a plaque was placed on the bridge to memorialize Ellis and his contributions.

FUNDER OF THE PUBLIC LAUNCH OF NETFLIX

William Bruce "Bill" Elmore earned bachelor's and master's degrees in electrical engineering in 1975 and 1976, respectively. After furthering his education with an MBA from Stanford University, Elmore embarked on a distinguished career. Beginning as president of Visual Engineering, he later transitioned to Inman and Bowman, a prominent venture capital firm, in 1987. There, he delved into Silicon Valley's technology landscape, investing in promising startups. In 1995, Elmore co-founded Foundation Capital, a pivotal player in nurturing companies like Netflix and Chegg. Now an emeritus partner, Elmore remains active in guiding emerging companies. Purdue recognition has included an Outstanding Electrical and Computer Engineer Award in 2003 and a Distinguished Engineering Alumnus Award in 2017. The Elmore Family School of Electrical and Computer Engineering also bears his name.

FIRST FEMALE CHIEF SCIENTIST OF THE U.S. AIR FORCE

Mica R. Endsley earned a master's degree in industrial engineering in 1985. After gaining experience at Texas Tech University and the Massachusetts Institute of Technology, Endsley founded SA Technologies in 1997. Her work at SA Technologies focused on cognitive engineering for advanced aviation, medical and military operating systems.

In 2013, Endsley made history by becoming the 34th chief scientist of the U.S. Air Force, the first woman and the first human factors engineer to hold the position. Endsley's Purdue honors include an Outstanding Industrial Engineer Award in 2013 and a Distinguished Engineering Alumna Award in 2023.

CEO AND CHAIRMAN OF TEXAS INSTRUMENTS

Thomas J. Engibous earned bachelor's and master's degrees in electrical engineering in 1975 and 1976, respectively, and received an honorary doctorate in 1997. After graduation, he began his career as a design engineer in the Texas Instruments semiconductor group.

Over two decades, he steadily rose through the ranks, being appointed president and CEO in 1996. Two years later, Engibous also assumed the role of chairman of the board. Under his leadership, Texas Instruments underwent a strategic transformation from a diversified conglomerate to a focused semiconductor company with a strong global presence. Purdue presented him with a Distinguished Engineering Alumnus Award in 1990.

INTEL'S EVP AND CHIEF GLOBAL OPERATIONS OFFICER

Keyvan Esfarjani earned bachelor's and master's degrees in industrial engineering in 1992 and 1994, respectively. His 28-year career at Intel Corporation commenced in 1996 when he joined the company as an industrial engineer. Esfarjani rose through the ranks, serving as a D2 plant manager in 2005 and later as general manager and director of Intel's California Tech Development.

In 2010, he assumed the role of co-chief executive officer at IM Flash Technologies, followed by appointments as corporate vice president of the Worldwide Memory Manufacturing Group and then senior vice president of manufacturing and operations at Intel.

In 2022, Esfarjani's exceptional leadership resulted in his promotion to executive vice president and chief global operations officer, responsible for Intel's worldwide manufacturing operations, strategic planning, quality assurance and supply chain while managing a workforce of 40,000 employees. Purdue presented Esfarjani with an Outstanding Industrial Engineer Award in 2022 and a Distinguished Engineering Alumnus Award in 2023.

CHAIRMAN AND CEO OF UPS

Michael L. "Mike" Eskew earned a bachelor's degree in industrial engineering in 1972. After graduation, he commenced his career at UPS as an industrial engineering manager. During the next 30 years, he steadily progressed through the ranks of the organization.

In 2002, Eskew was appointed chief executive officer and chairman of the board of UPS, overseeing its 360,000 global employees. While at UPS, Eskew was appointed to the President's Export Council and elected chairman of the US-China Business Council. In 2004, he was elected to the National Academy of Engineering. Currently, Eskew serves on the boards of 3M, IBM, and Eli Lilly and Company.

Ray Ewry was nicknamed "The Human Frog" by spectators at the Paris Olympic Games in 1900.

WINNER OF EIGHT INDIVIDUAL OLYMPIC GOLD MEDALS

Before swimmer Michael Phelps, track and field athlete **Ray Ewry** held the record as the U.S. Olympian with the most individual gold medals. A Purdue mechanical engineering graduate (bachelor's in 1894 and master's in 1897) and engineering faculty member, he had overcome childhood polio.

Ewry made history as the first Big Ten athlete to win Olympic gold—a feat he accomplished at the Paris Olympic Games in 1900, triumphing in the standing long jump, triple jump and standing high jump. By the end of the 1908 Games, Ewry had won eight gold medals, setting a medal count record that would last more than 100 years. Today, Purdue's Ray Ewry Sports Engineering Center bears his name.

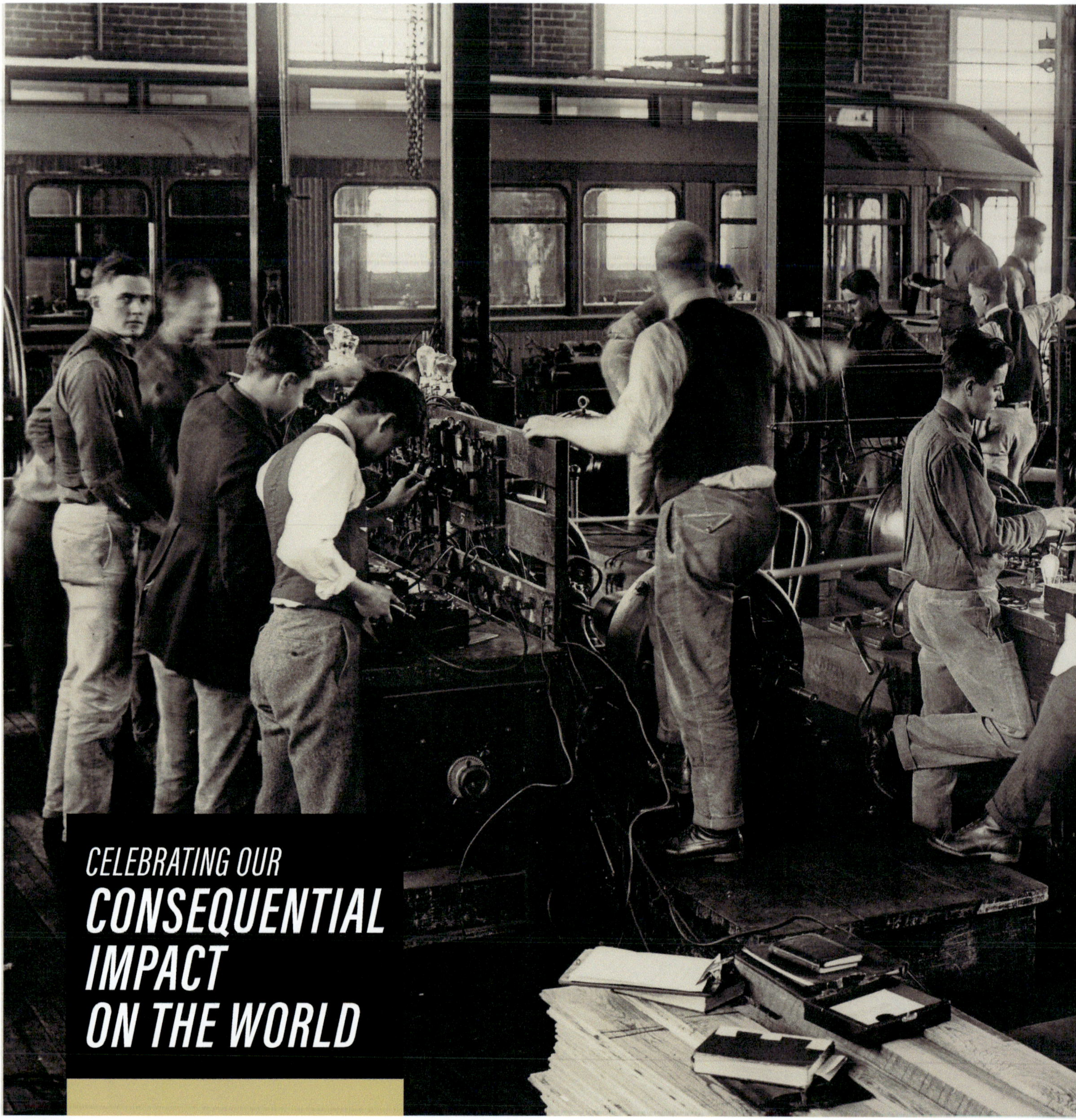

CELEBRATING OUR

CONSEQUENTIAL IMPACT ON THE WORLD

1923

Students work in a shop in the original Electrical Engineering Building. The "Louisiana" Interurban Car No. 2611 in the background was used for engineering testing by the 1904 St. Louis World's Fair agency. It was later rebuilt as an electric railway test car for Purdue.

"FATHER OF RADIO BROADCASTING"

Reginald Fessenden joined the faculty of Purdue's newly formed Electrical Engineering Department and served as its head from 1892 to 1893. Called the "Father of Radio Broadcasting," he is renowned for his pioneering work in wireless telegraphy. Fessenden's experiments initiated at Purdue led directly to wireless transmission of the human voice and music radio.

Breakthroughs included the first transmission of speech by radio, as well as the first two-way radiotelegraphic communication across the Atlantic Ocean. While at Purdue, Fessenden also played a key role in the installation of lighting for the Chicago World's Fair in 1893 in collaboration with the Westinghouse Corporation. The Institute of Radio Engineers awarded him an IRE Medal of Honor in 1921.

KEY CONTRIBUTOR TO DEVELOPMENT OF SEGWAY PT AND TESLA'S MODEL 3

J. Douglas "Doug" Field, a 1987 mechanical engineering bachelor's degree recipient, has blazed a career marked by innovation and leadership at the forefront of technology. At Segway, Field played a crucial role in creating the Segway PT, a groundbreaking self-balancing human transporter.

His contributions continued at Apple, where he served as vice president of special projects and helped develop several iconic consumer Mac products. Later, as Tesla's vice president of vehicle programs, Field led creation of the highly acclaimed Model 3 electric vehicle. Currently, at Ford Motor Company, he serves as chief EV, digital and design officer, spearheading development of Ford's EV portfolio. Purdue honored Field with a Distinguished Engineering Alumnus Award in 2015.

NVIDIA'S FIRST SALESPERSON —BEFORE IT HAD ANY PRODUCTS

Jeffrey "Jeff" Fisher earned a bachelor's degree in electrical engineering in 1980. Immediately after graduation, Fisher moved to Silicon Valley on the brink of its technology boom, and he has remained there ever since. Fisher's career began in the sales department of a tech company called Adaptec. Six years later, he transitioned to Weitex Corporation, where he assumed a sales leadership role. In 1994, Fisher made a significant career change by joining NVIDIA as the head of worldwide sales and the company's first salesperson.

At the time, NVIDIA employed approximately 25 people and had yet to ship its inaugural product. Since then, NVIDIA has evolved into a global leader in visual computing technologies. It is credited with inventing the GPU, a high-performance processor that generates interactive graphics across various platforms, including workstations, personal computers, game consoles and mobile devices. Today, one out of every three PCs and engineering workstations sold is powered by NVIDIA graphics, and the company's annual revenue exceeds $60 billion.

During his three decades with NVIDIA, Fisher has held several critical roles, such as executive vice president for worldwide sales and vice president of the graphics processing business unit. Currently, he serves as NVIDIA's senior vice president of the GeForce business unit. Purdue honored Fisher with an Outstanding Electrical and Computer Engineer Award in 2003 and a Distinguished Engineering Alumnus Award in 2009.

LEADER OF DUPONT TEAMS THAT COMMERCIALIZED POLYMERIC RESINS

Robert C. "Bob" Forney earned a bachelor's degree in chemical engineering (1947), a master's degree in industrial engineering (1948), and a PhD in chemical engineering (1950). He received an honorary doctorate from Purdue in 1981.

Upon graduation, he joined the DuPont Company, where he had a 39-year career, retiring as a senior vice president. At DuPont, he led engineering teams that commercialized Corfam, Corian, Dacron, Kevlar, Mylar, Nomex, Qiana and Tyvek. Forney Hall, the Chemical Engineering building at Purdue, is named in honor of Bob and his wife, Marilyn. Marilyn also earned a bachelor's degree in chemical engineering from Purdue.

FIRST WOMAN NAMED PRESIDENT OF THE AMERICAN SOCIETY OF CIVIL ENGINEERS

Patricia D. Galloway earned a bachelor's degree in civil engineering in 1978 after three years of study. Her career trajectory exemplified her exceptional talent and leadership in engineering. Beginning at CH2M Hill, a leading engineering firm, Galloway rapidly rose through the ranks, showcasing her expertise as a project engineer.

In 1981, she moved to The Nielsen-Wurster Group, where she became chief executive officer in 2001, marking a significant achievement for women leaders. In 2003, Galloway made history again as the first woman to be named CEO of the American Society of Civil Engineers, further solidifying her reputation as a trailblazer. Her tenure at the organization underscored her visionary leadership and dedication to advancing the engineering profession.

She subsequently served as CEO and was chairman of Global Pegasus Holdings, leaving a lasting impact on the industry. Purdue honored Galloway with a Distinguished Engineering Alumna Award in 1992.

CO-FOUNDER OF A $10B PUBLIC COMPANY

Ernest U. Gambaro earned bachelor's and master's degrees in aeronautics and astronautics in 1960 and 1961, respectively. He began his career as a systems engineer at The Aerospace Corporation, where he contributed to the design of the nation's premier orbital reconnaissance system.

After earning a law degree from Loyola Marymount University in 1975, Gambaro served as an attorney in Aerospace's corporate law department before becoming assistant general counsel for Computer Sciences Corporation in 1980.

In 1988, Gambaro co-founded Infonet Service Corporation, a pioneering venture that became the world's largest provider of secure data services. Under his leadership, Infonet experienced exponential growth, culminating in a landmark public offering on the New York Stock Exchange in 1999, valued at $10 billion. Gambaro's innovative contributions have had a lasting impact on the technology and telecommunications sectors.

Purdue recognized Gambaro with an Outstanding Aerospace Engineer Award in 2016 and a Distinguished Engineering Alumnus Award in 2021. In addition, his enduring commitment to education is honored through the Gambaro Graduate Program of Aeronautics and Astronautics.

FIRST WOMAN NAMED A SENIOR FELLOW AT ROLLS-ROYCE

Mary Lee Gambone graduated with a bachelor's degree in metallurgical engineering in 1982. After completing her undergraduate studies, she earned a master's degree at the Massachusetts Institute of Technology. Gambone commenced her career at General Motors' Allison Gas Turbine Division, leaving in 1990 to earn a PhD at the University of Virginia before becoming a team lead and materials engineer at the U.S. Air Force Research Laboratory, where she worked for seven years.

In 1998, Gambone transitioned to Rolls-Royce, joining its manufacturing technology group as a development engineer. Throughout her 26-year tenure at Rolls-Royce, she demonstrated exceptional leadership and expertise, holding key positions in the materials engineering and research and technology departments. Her contributions led to her promotion to head of materials engineering for North America and, in 2020, to global head of materials engineering.

In 2022, Gambone became the first woman in the company's history to attain the title of senior fellow. In addition to her Rolls-Royce accomplishments, Gambone is active in professional organizations. Currently, she serves as one of 11 directors of SAE International, contributing her expertise to shape the future of engineering practices and standards.

Purdue recognized Gambone with an Outstanding Materials Engineer Award in 2013, Cooperative Education Hall of Fame membership in 2015 and a Distinguished Engineering Alumna Award in 2024.

FOUNDER OF ONE OF AMERICA'S LARGEST HOMEBUILDERS

Roger B. Gatewood pursued a banking career after earning a bachelor's degree in mechanical engineering in 1968. However, in 1980, he ventured into entrepreneurship by founding an Illinois-based homebuilding company, Westfield Homes. The company experienced significant growth and expansion into Florida and the Carolinas, becoming one of the largest homebuilders in America.

Gatewood sold the business to Standard Pacific Homes in 2002. Purdue honored Gatewood with an Outstanding Mechanical Engineer Award in 2009 and a Philanthropist of the Year Award. The Roger B. Gatewood Wing of the Mechanical Engineering building is named for him.

INVENTOR OF BIOMEDICAL DEVICES USED ON HUNDREDS OF THOUSANDS OF PATIENTS

Leslie A. Geddes, the Showalter Distinguished Professor of Biomedical Engineering, arrived at Purdue in 1974 with a mission to help develop an organized biomedical engineering research center. That year, he became founding director of the Hillenbrand Biomedical Engineering Center. Geddes, an electrical engineer and physiologist, discovered the best places on the human chest for defibrillation or pacing—methods for restoring the proper rhythm of the heart.

He obtained more than 30 patents related to treating diseases of the heart and lungs, and he trained an entire generation of biomedical engineers. Inventions and discoveries over his 50-year career include the first clinical electromyograph (EMG), a respiratory monitoring system for NASA, an energy-efficient miniature defibrillator, a regenerative tissue graft made from pig intestines that has been used by surgeons on thousands of patients, and a pacemaker that automatically increases a patient's heart rate during exercise. He also invented a baby pacifier that delivers medicine.

Geddes' discoveries, licensed by such companies as Cook Biotech Inc., DePuy Inc., Eli Lilly and Company, and Hillenbrand, have generated tens of millions of dollars in royalties for Purdue, with one-third of those funds supporting other research to develop new technologies. Geddes officially retired from Purdue in 1991, when he was named a distinguished professor emeritus.

He received a National Medal of Technology and Innovation in 2006 and the Edison Medal from the Institute of Electrical and Electronics Engineers in 1994. Additionally, Geddes was inducted into the National Academy of Engineering in 1985 and received an Outstanding Contribution to Scientific Commercialization Award from Purdue in 2004.

INVENTOR OF THE FIRST ELECTRONIC TV RECEIVER AND FOUNDER OF INDIANA'S FIRST TV STATION

Roscoe H. George graduated with bachelor's and master's degrees in electrical engineering in 1922 and 1927, respectively. A prolific inventor, George obtained more than 30 patents. In 1929, he created the first electronic television receiver.

He established the first experimental television station in Indiana at Purdue under the call sign W9XG, with an FCC license granted in June 1930. And in 1940, he earned patents for his television system, a method for reviewing infrared systems, and a cathode ray tube. George was a professor in Purdue's electrical engineering program for more than 40 years, retiring in 1965.

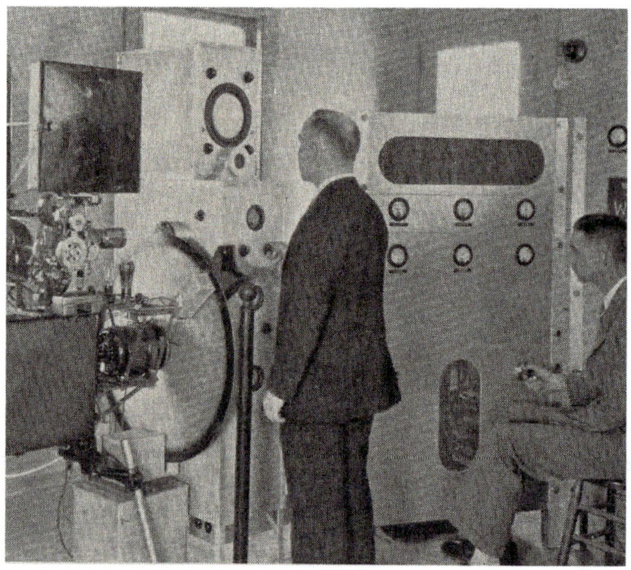

PURDUE UNIVERSITY ARCHIVES AND SPECIAL COLLECTIONS

THE MOST INFLUENTIAL PERSON WHEN IT COMES TO U.S. SPACE FLIGHT

William H. "Bill" Gerstenmaier received a bachelor's degree in aeronautics and astronautics in 1977 and an honorary doctorate in 2019. After completing his studies, he joined NASA's John Glenn Research Center, conducting research on supersonic tunnels, which contributed to information used for the reentry of the Space Shuttle.

In 1980, Gerstenmaier moved to the Johnson Space Center, where he participated in research concerning the Space Shuttle's propulsion and early stages of the design of the International Space Station (ISS). Gerstenmaier became heavily involved with the Space Shuttle program, eventually rising to Shuttle-Mir Program operations manager. He later was named program manager for the International Space Station.

From 2005 to 2019, Gerstenmaier served as NASA's associate administrator for human exploration and operations. In this role, he was described as arguably the most influential person when it comes to U.S. space flight. In 2020, Gerstenmaier completed his 40-year career with NASA to join SpaceX as a consultant. A year later, he became SpaceX's vice president of build and flight reliability.

Today, SpaceX is the leading launch services provider in the world, creating a fundamental shift in humanity's ability to access space—sending astronauts to the International Space Station and deploying critical scientific payloads for NASA to understand our universe better, as well as delivering Starlink satellites to orbit to provide high-speed internet for people all over the world, among other payloads for commercial and government customers.

Purdue recognized Gerstenmaier with an Outstanding Aerospace Engineer Award in 2003, a Distinguished Engineering Alumnus Award in 2007 and an Old Masters induction in 2008.

BLIND BOILERMAKER WHO WAS CEO OF GOODWILL INDUSTRIES INTERNATIONAL INC.

James D. "Jim" Gibbons, who earned a bachelor's degree in industrial engineering in 1985, faced the challenge of losing his sight during his early years at Purdue, ultimately becoming completely blind. With the support of many on campus, Gibbons persevered and successfully completed his degree requirements.

Following graduation, he joined AT&T. While working there, he became the first blind person to earn an MBA from Harvard University. Gibbons spent 13 years at AT&T, rising to CEO of the campuswide access solutions business unit. In 1998, he transitioned to CEO of the National Institute for the Blind, where he served for a decade.

Subsequently, Gibbons was CEO of Goodwill Industries International Inc., supervising operations of this prominent nonprofit organization from 2008 to 2018. Purdue honored Gibbons with a Distinguished Engineering Alumnus Award in 2014.

THE U.S. SENATE'S SERGEANT AT ARMS

The U.S. Senate's Sergeant at Arms, **Karen Gibson**, graduated with a bachelor's degree in industrial engineering in 1986. Her academic journey continued with master's degrees in national security strategy from the National War College and in strategic intelligence from the National Intelligence University. Gibson recently concluded a 33-year military career, achieving the rank of lieutenant general in the U.S. Army. She excelled in various intelligence and cyberspace operations roles throughout her extensive service.

Gibson's Army career culminated in her appointment as a deputy director of national intelligence. In previous assignments, Gibson held crucial positions, such as director of intelligence for United States Central Command, as well as director of intelligence for Combined Joint Task Force—Operation Inherent Resolve, the multinational coalition dedicated to defeating the ISIS caliphate in Iraq and Syria. She also served as deputy commanding general for U.S. Army Cyber Command.

A seasoned combat veteran, Gibson demonstrated exemplary leadership in intelligence-operations fusion centers across Iraq, Afghanistan, Korea, East Africa and the U.S. In March 2021, Gibson made history by becoming only the second woman appointed as the U.S. Senate's Sergeant at Arms. In this pivotal role, she is entrusted with maintaining security in the Capitol and all Senate buildings, and with ensuring the protection of members of Congress.

As the Senate's chief law enforcement officer, Gibson oversees a range of responsibilities, including managing computers and technology support services, escorting the president and other heads of state or official guests of the Senate in the Capitol, and holding custody of the Senate gavel.

AMERICA'S "FIRST LADY OF ENGINEERING"

Lillian Gilbreth, a professor of management in the College of Engineering from 1935 to 1948, was the first woman to serve as a Purdue engineering professor and became known as America's "First Lady of Engineering." A pioneer in industrial and organizational psychology, Gilbreth made significant contributions to the field. She played a crucial role in designing the modern kitchen, introducing the "work triangle" concept for optimal efficiency.

Additionally, Gilbreth invented several household conveniences, including the foot-pedal trash can, shelving inside refrigerators and wall light switches. Her remarkable life and achievements were immortalized in the books *Cheaper by the Dozen* and *Belles on Their Toes*, written by two of her 11 children and adapted into movies.

In 2018, Purdue established the Lillian Gilbreth Postdoctoral Fellowships program to prepare recently awarded PhD recipients for careers in engineering academia.

ONE OF THE FIRST AFRICAN AMERICAN PIONEERING TECHNOLOGISTS IN SILICON VALLEY

Frank S. Greene Jr. earned a master's degree in electrical engineering in 1962. After graduating, Greene entered the U.S. Air Force, where he achieved the rank of captain during his four years of service. Following his military service, he joined the research and development department of Fairchild Semiconductor. While working, Greene earned a PhD from Santa Clara University in 1970.

He excelled as a semiconductor researcher, entrepreneur, teacher and venture capitalist. Greene was awarded the patent for an integrated circuit that was the world's fastest microchip at the time and contributed to Fairchild Semiconductor's leadership in the semiconductor industry in the late 1960s.

He taught electrical engineering and computer science at five universities, and he founded two software companies, one of which *Black Enterprise* recognized among the top 100 Black-owned businesses. Additionally, Greene co-founded New Vista Capital, a venture capital firm that made early-stage investments in information technology companies owned by women and minorities.

Purdue presented Greene with an Outstanding Electrical and Computer Engineer Award in 1999 and a Distinguished Engineering Alumnus Award in 2002. The Frank S. Greene Jr. Middle School in Palo Alto, California, was named in his honor in 2018.

INNOVATOR WHOSE INVENTIONS HAVE IMPROVED THE LIVES OF MILLIONS WORLDWIDE

James F. Greenleaf earned a master's degree in engineering science from Purdue in 1966, as well as a joint PhD in engineering science from the Purdue College of Engineering and the Mayo Graduate School of Medicine in 1970. Following his studies, Greenleaf accepted a research position at the Mayo Graduate School of Medicine, where he remained, and he joined the Mayo Medical School faculty in 1973. During his Mayo career, Greenleaf's roles have been diverse. He initially served as a faculty member, teaching various programs, before becoming a professor of biomedical engineering and associate professor of medicine in 1983. Concurrently, since 1974, he has been actively engaged in Mayo's Ultrasound Research Laboratory. Greenleaf's contributions to biomedical engineering are extensive and highly regarded. He has garnered many national awards from professional societies; published more than 450 peer-reviewed articles; edited or authored five books in the field; and obtained 51 patents, several of which have gained international recognition. His prolific research career spans areas including development of lifesaving devices. Beyond his individual accomplishments, Greenleaf's impact extends to the education of numerous graduate students at the Mayo Clinic and Mayo Clinic School of Medicine, influencing the next generation of biomedical engineers and medical professionals. Purdue presented Greenleaf with a Distinguished Engineering Alumnus Award in 2017.

FOUNDER OF A COMPANY THAT CREATES CANCER-KILLING RADIOISOTOPES

Terry Grimm graduated with a bachelor's degree in nuclear engineering. He later earned a PhD in nuclear engineering and plasma physics from the Massachusetts Institute of Technology. Afterward, Grimm worked as a physicist at the U.S. Department of Energy's Superconducting Super Collider laboratory in Dallas.

In 1994, he became a senior physicist at the National Superconducting Cyclotron Laboratory at Michigan State University. In 2005, Grimm founded Niowave Inc., focusing on commercializing superconducting electron linear accelerators. Under his leadership, Niowave became a global frontrunner in superconducting accelerator research. Grimm's pioneering work earned Niowave recognition as the second U.S. vendor for accelerating cavities for the International Linear Collider in 2008. The following year, Niowave became the world's first private company to design, fabricate and test a superconducting electron injector.

In 2019, Grimm received a $15 million federal grant to support Niowave's production of a key medical radioisotope for treating heart disease and cancer, and Niowave is actively involved in cancer clinical trials. Purdue honored Grimm with a Distinguished Engineering Alumnus Award in 2019.

SECOND AMERICAN IN SPACE

Virgil "Gus" Grissom earned a bachelor's degree in mechanical engineering in 1950. After graduating, Grissom joined the U.S. Air Force, where he served as a pilot and flew 100 combat missions during the Korean War, earning the Distinguished Flying Cross for bravery. In 1959, Grissom was selected as one of the seven astronauts for NASA's Project Mercury, a critical initiative aimed at surpassing Russia in the space race.

On July 21, 1961, he became the second American to venture into space, piloting the Mercury-Redstone 4 mission. Nearly six years later, Grissom lost his life in a fire that erupted inside the cockpit of Apollo 1 during a prelaunch test. Grissom's contributions to space exploration were widely recognized after his death. He was inducted into the Astronaut Hall of Fame, Space Hall of Fame and Aviation Hall of Fame. Additionally, he posthumously was awarded a Presidential Medal of Freedom and a NASA Distinguished Service Medal for his extraordinary service and sacrifices. Purdue paid tribute to Grissom's legacy by naming its School of Industrial Engineering building, Grissom Hall, in his honor.

NASA

DEVELOPER OF 17M-PLUS SQUARE FEET OF OFFICE SPACE—AND MORE

Marc Lee Hagle earned a bachelor's degree in electrical engineering in 1971 and a master's degree in business and industrial administration in 1972. His career has reshaped the landscape of real estate development. Inspired by Walt Disney's ventures in Florida, Hagle crafted his business plan, laying the foundation for Tricor International.

Since the company's inception, its subsidiaries have left an indelible mark across the nation, spearheading construction of diverse projects ranging from single-family homes to multi-use communities, medical facilities to recreational centers and office complexes to warehouses. Their footprint spans Alaska to West Virginia, covering states including California, Texas and Florida.

With a portfolio of more than 17 million square feet of commercial space and thousands of residential units, Tricor International has solidified its position as a leader in the real estate sector. Today at Purdue, Marc and Sharon Hagle Hall is home to the "All-American" Marching Band and other musical ensembles. Additionally, Hagle has contributed by sharing his wealth of knowledge with future generations as an adjunct professor in Purdue's Mitch Daniels School of Business.

NASA

DEPUTY MANAGER OF AMERICA'S SPACE SHUTTLE PROGRAM

N. Wayne Hale Jr. earned his master's degree in mechanical engineering in 1978. Following graduation, Hale joined NASA as a propulsion officer at the Johnson Space Center. Over the next decade, he advanced through the ranks, eventually serving as a flight director in Mission Control for 41 Space Shuttle missions. Later, Hale relocated to the Kennedy Space Center, where he was deputy manager of the Space Shuttle program, responsible for overall management, integration and operation.

Throughout his career, Hale received numerous honors, including NASA's Outstanding Leadership Medal and Exceptional Service Medal. Purdue presented Hale with a Distinguished Engineering Alumnus Award in 2011.

ONE OF THE TOP VENTURE CAPITALISTS IN AMERICA

Mamoon Hamid graduated with a bachelor's degree in electrical engineering in 1997. He commenced his Silicon Valley career at Xilinx, where he worked in engineering and marketing roles before transitioning to venture capital. In 2005, he joined U.S. Venture Partners and spent six years there before co-founding a company called Social Capital.

At Social Capital, he played a key role in helping companies like Intercom, Greenhouse and Netskope get started, and he was the first outside investor in Slack. In 2017, Hamid moved to Kleiner Perkins, one of Silicon Valley's most renowned venture firms, where he became a managing member and general partner.

He has been recognized by *The New York Times* as one of the Top 100 Venture Capitalists in the country and has appeared multiple times on *Forbes'* Midas List for top tech investors. Purdue presented Hamid with an Outstanding Electrical and Computer Engineer Award in 2020.

SECOND ONLY TO THOMAS EDISON IN U.S. PATENTS

Clarence Weston Hansell earned a bachelor's degree in electrical engineering in 1919. After graduation, he began his career at General Electric, where he led factory testing of high-frequency generators. Shortly afterward, he transferred to GE's Radio Engineering Department to work on testing transoceanic radio transmitting stations. Later, Hansell moved to the Radio Corporation of America, where he played a pivotal role in developing its transmitting facilities and founded the RCA Transmission Laboratory.

In 1932, he became particularly interested in ionized air and its biological effects on people. Upon retiring from RCA in 1963, Hansell held more than 300 U.S. patents, making him the second-most prolific inventor in American history, second only to Thomas Edison. Purdue awarded him an honorary doctorate in electrical engineering in 1952.

CO-FOUNDER OF THE NATIONAL SOCIETY OF BLACK ENGINEERS

Anthony "Tony" Harris, who earned a bachelor's degree in mechanical engineering in 1975 and an honorary doctorate in 2013, played a crucial role in founding the National Society of Black Engineers (NSBE).

As president of Purdue's chapter of the Society of Black Engineers (SBE) during his senior year, Harris, along with five friends from Chicago, known as the "Chicago Six," supported by Purdue President Arthur G. Hansen, expanded the group nationally. In April 1975, they invited universities nationwide to a meeting at Purdue, where NSBE was formed. Today, NSBE is one of the largest student-run organizations in the U.S., with more than 31,000 members.

Harris began his career as a design engineer at Standard Oil Company before earning an MBA. He held significant roles at Ford Motor Company, Pacific Gas & Electric and Calpine Corporation. In 2007, he became president and CEO of Campbell/Harris Security Equipment Company. Purdue recognized him with an Outstanding Mechanical Engineer Award in 1999 and a Distinguished Engineering Alumnus Award in 2008.

ONE OF SILICON VALLEY'S TOP INNOVATORS AND ENTREPRENEURS

Gerald D. "Jerry" Held earned a bachelor's degree in electrical engineering in 1970 and spent six formative years at David Sarnoff Laboratories. His subsequent journey led to Silicon Valley, where he carved out one of the most successful technology careers in the industry. Held's expertise spans various domains. He has profoundly shaped computing and database technology, playing a central role in creating groundbreaking database innovations and establishing new companies.

His pioneering work has included development of the INGRES database while at UC Berkeley, followed by his instrumental contribution to Tandem's growth from a startup to a company with more than $2 billion in annual revenue. At Oracle, Held steered the server products division during a period of extraordinary growth, overseeing a surge in annual revenue from $1.5 billion to $6 billion.

His leadership extended beyond individual companies, as he served as "CEO-in-residence" at Kleiner Perkins, providing strategic guidance and mentorship to emerging ventures. Held's influence in the technology sector is further underscored by his roles as lead director of Business Objects until its acquisition by SAP and as executive chairman of Vertica until its purchase by Hewlett-Packard.

Currently, he serves as chairman of Tamr and sits on the boards of Informatica, NetApp, Kalio, MemSQL and Copia Global. Purdue honored Held with an Outstanding Electrical and Computer Engineer Award in 1997 and a Distinguished Engineering Alumnus Award in 2001.

HIGHEST-RANKING WOMAN IN THE FBI IN 2018

Amy S. Hess earned a bachelor's degree in aeronautics and astronautics in 1989. She began her career at the FBI in 1991 as a special agent and dedicated 29 years to the organization.

Throughout her tenure, Hess held various notable roles, including a deployment to Afghanistan to support counterterrorism operations, as well as leadership positions in the International Operations Division; National Security Branch, Operational Technology Division; and Science and Technology Branch. She was appointed special agent in charge of the Memphis, Tennessee, and Louisville, Kentucky, field offices.

In 2018, Hess reached a major milestone when she was appointed executive assistant director of the FBI's Criminal, Cyber, Response and Services Branch, becoming the highest-ranking woman in the organization at the time. Purdue recognized her with a Distinguished Engineering Alumna Award in 2017.

CELEBRATING OUR CONSEQUENTIAL IMPACT ON THE WORLD

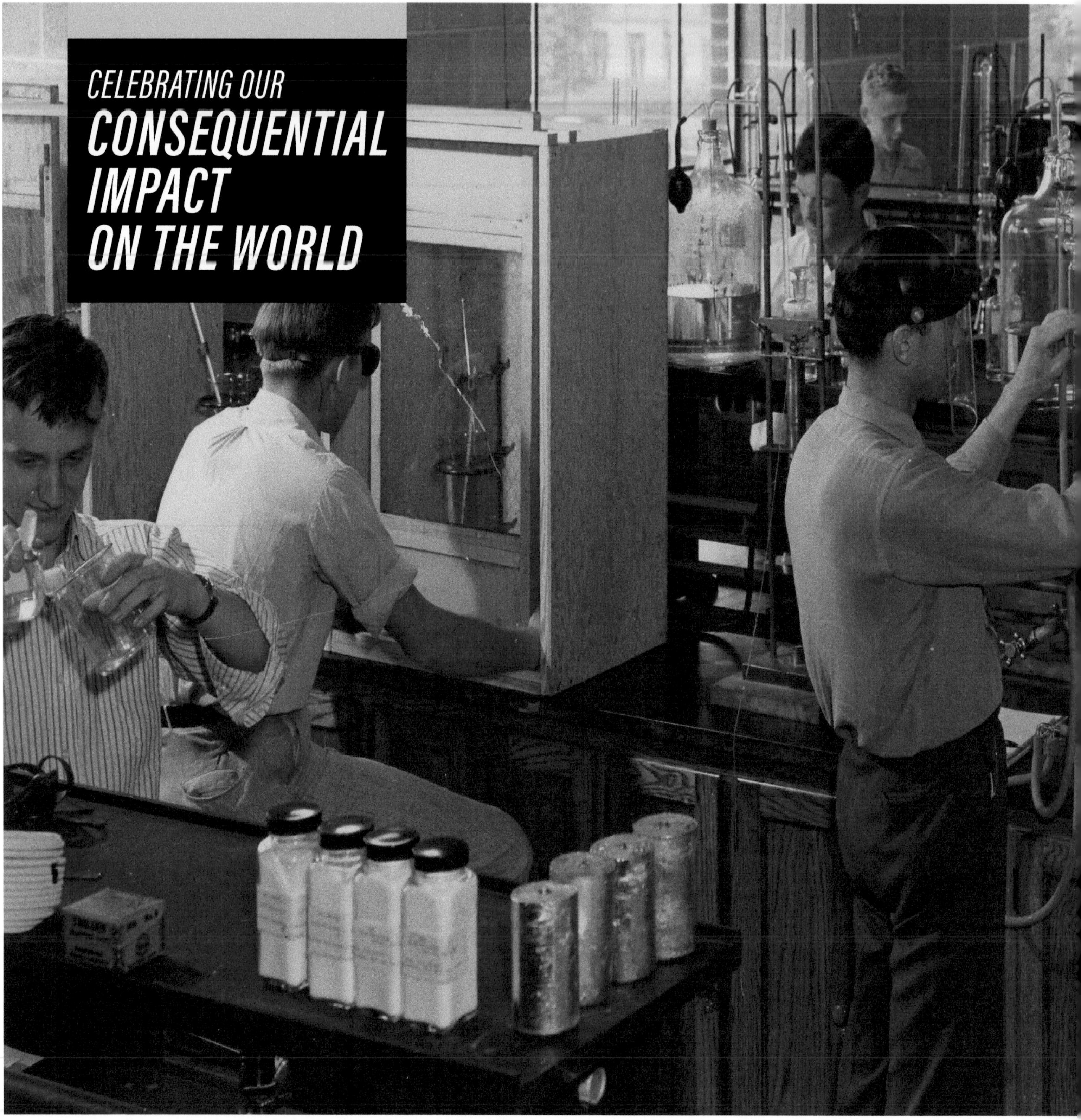

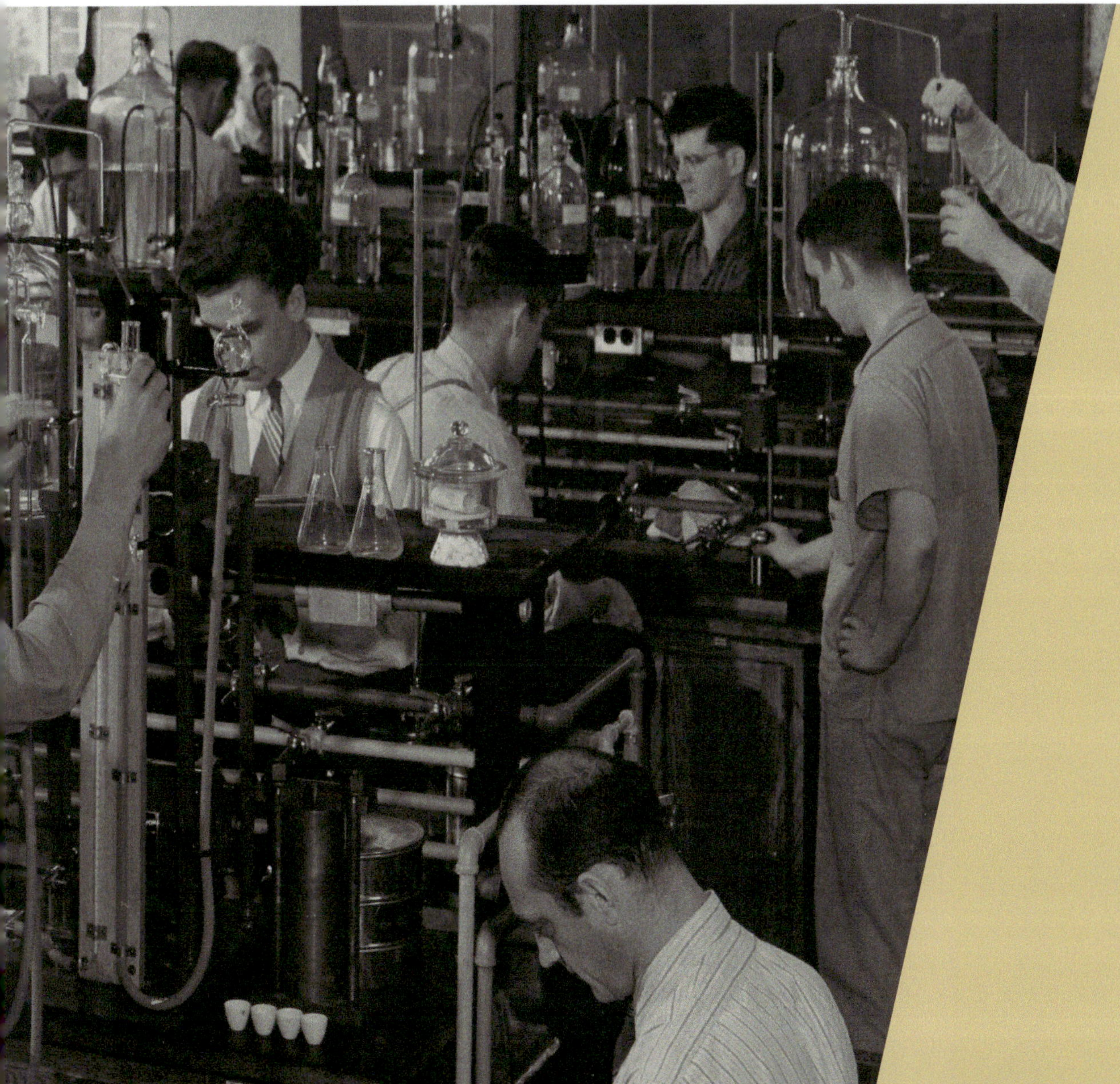

1936

Students work in a chemical engineering laboratory. This predates the current Chemical Engineering building, constructed in 1939 and renamed Forney Hall in 2004.

A PIONEER OF SOLAR ENERGY

Walter J. Hesse obtained a bachelor's (1944), a master's (1948), and a PhD (1951) in mechanical engineering. His thirst for knowledge led him to the U.S. Naval Academy, UC Berkeley and Sandia National Laboratories. During World War II, Hesse served as an assistant engineer aboard the submarine USS Blenny. Postwar, he became chief engineer at the Test Pilot School, U.S. Naval Air Test Center, where he mentored future NASA astronauts. Hesse's career flourished in senior management roles, notably as vice president at E-Systems' Energy Technology Center.

A pioneer in concentrated solar photovoltaics and thermal technology, Hesse installed a groundbreaking hybrid PV electrical/solar thermal system at Dallas Fort Worth International Airport in 1982. He founded Entech Solar the following year. Purdue recognized Hesse with a Distinguished Engineering Alumnus Award in 1966, an Outstanding Mechanical Engineer Award in 1991 and an Outstanding Aerospace Engineer Award in 1999. Today, the Walter Hesse Scholarship supports Purdue engineering students.

ONE OF THE "50 MOST IMPORTANT WOMEN IN SCIENCE"

Kathleen Howell is the Hsu Lo Distinguished Professor of Aeronautics and Astronautics. Alongside one of her Purdue graduate students, Howell pioneered a groundbreaking approach to powering spacecraft to their destinations, harnessing the untapped free energy within the solar system. The innovative trajectory design significantly enhanced the efficiency of NASA's low-fuel Genesis probe, launched in 2001. By leveraging the gravity fields of various celestial bodies, Howell and her student identified natural pathways that spacecraft could exploit, reducing the fuel required for interplanetary travel.

This pioneering finding earned Howell a place on *Discover*'s list of the "50 Most Important Women in Science" in 2022. She also is a distinguished member of the National Academy of Engineering. In addition to her pivotal research, Howell has demonstrated a deep commitment to educating and inspiring future generations of engineers. Purdue honored her with the A.A. Potter Award for Excellence in Teaching in 1984 and 2011. In 2025, she received the Faculty Excellence Award for Graduate Student Mentorship.

ONLY WOMAN TO LEAD A U.S. NUCLEAR WEAPONS LABORATORY

Jill M. Hruby graduated with a bachelor's degree in mechanical engineering in 1981. For 36 years, she was based at Sandia National Laboratories, a research facility of the U.S. Department of Energy's National Nuclear Security Administration. Hruby began as a scientist and rose to become Sandia's director, making her the only woman to lead a U.S. nuclear weapons laboratory.

Directing a team of 12,000 scientists, engineers and contractors, she addressed critical security challenges. Following her retirement in 2017, Hruby was the inaugural Sam Nunn Distinguished Fellow at the Nuclear Threat Initiative from 2018 to 2019. She also served on several advisory committees, including the Defense Science Board and the National Academy of Science Committee for International Security and Arms Control.

On July 26, 2021, Hruby was sworn in as the DOE's undersecretary for nuclear security and administrator of the NNSA. In this role, she leads the NNSA to advance the nation's nuclear security missions, ensuring that the Nuclear Security Enterprise employs cutting-edge science to maintain the stockpile without testing, develops new technologies to reduce costs and enhance security, and stays ahead of adversaries. Purdue honored Hruby with a Distinguished Engineering Alumna Award in 2014.

CHAIRMAN AND CEO OF AMERICA'S LARGEST STEEL MANUFACTURER

F. Kenneth "Ken" Iverson earned a master's degree in mechanical engineering in 1947 and received an honorary doctorate in 1988. He began his career as a research physicist at International Harvester. Five years later, he transitioned to Illium Corporation, where he served as a chief engineer. Over the next eight years, Iverson held positions at Indiana Steel Products, Cannon-Muskegon and Coast Metals Company, where he established a spectrographic laboratory, served as chief metallurgist and was executive vice president, respectively.

In 1961, Iverson joined the Nuclear Corporation of America as its general manager. When the company faced bankruptcy five years later, Iverson, heading the only profitable division, was appointed chairman and CEO. Under his leadership, the company underwent significant transformations, including racial integration. Rebranded as Nucor in 1972, the company emerged as America's largest and most prosperous steel producer.

Iverson received a National Medal of Technology and Innovation in 1991. Three years later, he was inducted into the National Academy of Engineering, cementing his legacy as a pioneering figure who transformed the steel manufacturing landscape. Purdue presented Iverson with an Outstanding Mechanical Engineer Award in 1991.

INVENTOR OF TECHNOLOGY CRITICAL TO NATIONAL AND GLOBAL SECURITY

Charles V. "Jack" Jakowatz Jr. earned three Purdue electrical engineering degrees: a bachelor's degree (1972), a master's degree (1973) and a PhD (1976). Immediately after graduation, he embarked on an engineering career at Sandia National Laboratories in New Mexico, where he remained for 25 years.

While at Sandia, Jakowatz and his team focused on developing predominantly classified technology for the U.S. government that has been deemed crucial to national security domestically and globally. His contributions include advancement of synthetic aperture radar (SAR), an esoteric tool for remote sensing and observations that is considered among the most significant national security breakthroughs. Due to the sensitive nature of his work, many of Jakowatz's accomplishments remain classified.

The National Academy of Engineering elected him a member in 2003, citing his groundbreaking innovations in SAR-imaging processing, which are critical for military applications and environmental monitoring. Purdue honored him with an Outstanding Electrical and Computer Engineer Award in 2002 and a Distinguished Engineering Alumnus Award in 2005.

PURDUE'S HEAD FOOTBALL COACH AND HEAD BASKETBALL COACH—AT THE SAME TIME

Alpha P. Jamison earned his bachelor's (1895) and master's (1897) degrees in mechanical engineering. While a student, Jamison was celebrated as the star quarterback and captain of the Purdue football team, and for many years, he was placed on the "all-time team" as the greatest QB in Purdue's history.

After graduation, he was an assistant professor of mechanical drawing at Purdue for 18 years. Additionally, Jamison held coaching positions in Purdue football and basketball. He served as head football coach for three seasons (1898 to 1900) and as head basketball coach from 1899 to 1901.

Interestingly, despite his football prowess, Jamison achieved greater success as a head basketball coach, with a record of 12 wins and one loss in basketball, compared with 11 wins, 11 losses and one tie in football.

PURDUE UNIV. ARCHIVES AND SPECIAL COLLECTIONS

Purdue football team, 1889. From the 1900 Debris *yearbook.*

CONTRIBUTOR TO MANY OF SILICON VALLEY'S BIGGEST SUCCESS STORIES

Rouz Jazayeri earned bachelor's and master's degrees in electrical engineering in 2002 and 2003, respectively. He began his career at Intel Corporation as a chip design engineer, contributing to the company's development of its first 45nm processor. He later joined Kleiner Perkins, a renowned venture capital firm in Silicon Valley, where he served as chief of staff to John Doerr.

At Kleiner Perkins, Jazayeri drove early-stage hard-tech investments and supported the growth of notable startups, including Ring (acquired by Amazon), LuxVue (acquired by Apple), MyFitnessPal (acquired by Under Armour) and QuantumScape.

In 2017, Jazayeri co-founded Catapult Ventures, a venture capital firm dedicated to supporting emerging startups. Through Catapult Ventures, he has funded and supported more than 20 innovative startups, including Catapult portfolio company Oculii (acquired by Ambarella) and substantially enhancing the entrepreneurial ecosystem.

DEVELOPER OF THE "STABILIZED APPROACH" –THE WORLD STANDARD FOR PILOTS

Ralph Johnson graduated from Purdue's mechanical engineering program in 1930, and he received an honorary doctorate in 2008 at age 101. Following graduation, he began pilot training at March Field in California, spending several adventurous years with the U.S. Army Air Corps before transitioning to civilian life.

Johnson made substantial contributions to air safety, pioneering hallmark designs and innovations. He dedicated his life to enhancing existing technology and creating new procedures, notably developing the standardized procedure for safely landing large aircraft, known as the "stabilized approach," which revolutionized flight and remains standard practice for all pilots.

After a distinguished career that included more than 7,000 test flights as the chief test pilot for United Airlines, Johnson ventured into politics, serving two terms in the Wyoming House of Representatives in the 1950s. Despite his political engagements, Johnson continued pursuing his passion for flying, owning a squadron of 22 World War II aircraft and participating in missions such as bombing fire ants and fighting forest fires until he was 82. Recognized as an aviation pioneer, he was inducted into the Wyoming and Arizona Aviation Halls of Fame. Johnson lived to be 103.

PRESIDENT OF RAYTHEON'S SPACE AND AIRBORNE SYSTEMS —WITH 13,000 EMPLOYEES

Jack R. Kelble received a bachelor's degree in electrical engineering in 1965. He embarked on his career at RCA, where he dedicated the first 14 years to hardware design, radar system engineering and various engineering projects. In 1979, Kelble transitioned to Raytheon, where he spent the next 26 years leading teams, divisions and companies within the corporation. His final role, as president of Space and Airborne Systems, positioned him at the helm of Raytheon's largest division, with 13,000 employees and annual revenue of $4 billion from defense systems. Purdue awarded him an honorary doctorate in 2006.

WORLD RECORD SETTER FOR FLYING AN AIRPLANE AT 126,200 FEET ABOVE EARTH

Iven C. "Kinch" Kincheloe Jr. graduated with a bachelor's degree in aeronautical engineering in 1949. A Reserve Officers' Training Corps (ROTC) student at Purdue, he joined the U.S. Air Force's flight training program upon graduation. Kincheloe distinguished himself during the Korean War, flying 131 combat missions and earning a Silver Star for downing five MIG-15s.

Following his military service, Kincheloe became a test pilot for the Bell X-2 program. In one historic flight, he reached speeds exceeding 2,000 miles per hour and soared to 126,200 feet, marking the first time an aircraft had flown higher than 100,000 feet. This accomplishment earned him the nickname "America's No. 1 Spaceman" although the flight was suborbital and did not meet the threshold of space. Kincheloe Air Force Base in Michigan's Upper Peninsula was named after him.

CO-INVENTOR OF PFIZER'S COMMERCIAL DRUG ZMAX

Richard W. Korsmeyer, a co-inventor of Pfizer's commercial drug Zmax, received two Purdue degrees in chemical engineering: a master's degree and a PhD in 1980 and 1983, respectively. Korsmeyer started his career at BFGoodrich as a senior research and development engineer before joining Pfizer in 1986.

Over the next 34 years, he played a crucial role in developing and commercializing numerous pharmaceutical technologies. One of Korsmeyer's notable contributions is his collaboration with former Purdue engineering professor Nicholas Peppas in devising the exponential equation (known as the Korsmeyer Equation) for quantifying controlled drug release from polymeric matrices. Published in 1983, this work is the most-cited paper ever published by the *International Journal of Pharmaceutics*.

Among his achievements at Pfizer, Korsmeyer co-invented Zmax, a sustained-release product that delivers a full course of the antibiotic azithromycin orally in a single dose. Azithromycin has become one of the most prescribed medications in the U.S. Korsmeyer's expertise and impact extend beyond his inventions.

He holds memberships in such organizations as the National Academy of Engineering and the Connecticut Academy of Science and Engineering. Korsmeyer also is a fellow of the American Institute of Chemical Engineers and the American Institute of Medical and Biological Engineering. Purdue honored him with a Distinguished Engineering Alumnus Award in 2014.

DESIGNER AND DEVELOPER OF A MOBILE APP ACQUIRED BY LINKEDIN

Akshay D. Kothari received a bachelor's degree in mechanical engineering in 2007. Following graduation, he embarked on a multifaceted career. Initially, Kothari joined SDL Ventures as an investment analyst, but soon afterward, he pursued further education, earning a master's degree in electrical engineering from Stanford University. Next, he delved into mobile app development, where his ingenuity and creativity flourished.

One of his notable creations, Pulse News, experienced exponential growth, garnering a user base of 30 million within three years. This success captured the attention of Apple CEO Steve Jobs, who showcased Kothari's app during his keynote address in 2010. Three years later, Kothari and his business partner successfully negotiated the sale of Pulse to LinkedIn for $90 million. Subsequently, Kothari transitioned to LinkedIn, becoming vice president of product, as well as head of LinkedIn India, where he played a pivotal role in advancing the platform's growth in a key market.

In 2018, Kothari started a new career chapter by joining Notion as chief operating officer. His continued dedication and innovative contributions have earned him widespread attention. Pulse was inducted into Apple's App Store Hall of Fame, and it received an Apple Design Award and recognition as one of *Time*'s Top 50 iPhone Apps of 2011. Purdue honored Kothari with an Outstanding Electrical and Computer Engineer Award in 2014 and a Distinguished Engineering Alumnus Award in 2021.

CHAIRMAN AND CEO OF TWO $40B COMPANIES

Keith J. Krach earned a bachelor's degree in industrial engineering in 1979 and received an honorary doctorate in 2018. Upon graduation, he joined General Motors, where he quickly became the youngest vice president in company history, at 26. While at GM, Krach co-founded GMF Robotics, which later became the world's largest supplier of industrial robots.

In 1987, Krach left GM to pursue opportunities in Silicon Valley. There, he co-founded Ariba and served as the company's CEO and chairman from 1996 to 2003. Under his leadership, Ariba surpassed a market capitalization of $40 billion in 2000. Krach later served as chairman and CEO of DocuSign, during which time the company exceeded the $40 billion market cap milestone.

Krach also has been actively involved in academia and public service. He was chairman of Purdue's Board of Trustees from 2007 to 2013. In 2019, the U.S. Senate confirmed his appointment as undersecretary of state for economic growth, energy and the environment. He currently is chairman of the Krach Institute for Tech Diplomacy at Purdue, which he co-founded.

His achievements have received wide recognition, including a nomination for the Nobel Peace Prize in 2022. Purdue presented Krach with an Outstanding Industrial Engineer Award in 2000 and a Distinguished Engineering Alumnus Award in 2006.

CHAIRMAN OF A WORLD LEADER IN DIGITAL DISPLAY TECHNOLOGY

Aelred "Al" J. Kurtenbach earned a doctorate in electrical engineering in 1968. While at Purdue, Kurtenbach decided to pursue entrepreneurship. The year he obtained his PhD, he co-founded and became chairman of Daktronics. Specializing in manufacturing electronic scoreboards, Daktronics quickly emerged as a global leader in digital scoreboard technology. Today, its digital displays are in stadiums, arenas and corporate settings worldwide.

Kurtenbach also has taught electrical engineering at Purdue and South Dakota State University. Purdue honored him with an Outstanding Electrical Engineer Award 2001 and a Distinguished Engineering Alumnus Award in 2006.

PRESIDENT AND CEO OF A NEARLY $5B GLOBAL COMPANY

Richard G. Kyle graduated with a bachelor's degree in mechanical engineering in 1988. He currently serves as president, chief executive officer and a board member of The Timken Company, a renowned global manufacturer of bearings and power transmission products that he joined in 2006.

Kyle's earlier career included management positions at Cooper Industries and Hubbell Inc. At Timken, Kyle has held various leadership roles, such as chief operating officer of the bearings and power transmissions group. He also headed Timken's aerospace, steel and mobile industries business segments. Under Kyle's leadership, Timken has grown significantly, with almost $5 billion in global revenue, as well as more than 19,000 employees operating in 46 countries. Purdue presented Kyle with a Distinguished Engineering Alumnus Award in 2018.

U.S. AIR FORCE'S "FATHER OF LASERS"

Retired Maj. Gen. **Donald L. Lamberson** began his distinguished 35-year U.S. Air Force career as a Reserve Officers' Training Corps (ROTC) student at Purdue, where he earned a bachelor's degree in chemical engineering in 1953.

Recognized for his expertise in directed energy physics, Lamberson was known as the "Father of Lasers" in the Air Force. During the 1970s and 1980s, Lamberson played a central role in advancing laser technology while heading the Air Force's Weapons Laboratory. Under his leadership, the laboratory developed the first high-powered laser capable of precision targeting from an aircraft.

Lamberson retired from the Air Force in 1989. Throughout his service, he received many accolades, including a Distinguished Service Medal, a Defense Superior Service Medal, a Legion of Merit and an Air Force Commendation Medal. Purdue recognized Lamberson with a Distinguished Engineering Alumnus Award in 2002.

MANAGER OF MORE THAN ONE-THIRD OF THE WORLD'S SECURITIES

Peter K. Lambertus earned a bachelor's degree in electrical engineering in 1967. Following graduation, he served as an officer in the U.S. Navy's nuclear submarine force for five years. Lambertus then joined Stone & Webster Engineering, where he worked as a nuclear systems engineer and later became manager of computer systems development.

In 1984, Lambertus founded Charles River Development, a financial services company that eventually managed more than one-third of the world's assets. He led the company until its acquisition by State Street Corporation in 2018 for $2.6 billion.

Purdue recognized Lambertus with a Distinguished Engineering Alumnus Award in 2020. Additionally, Lambertus Hall in Purdue's Gateway Complex is named after him and his wife, Ann Lambertus.

APPLE'S VP OF OPERATIONS

Rita S. Lane earned a master's degree in electrical engineering in 1991, after graduating from the U.S. Air Force Academy in 1984. Her successful career started at IBM, where she spent 15 years and ultimately served as vice president for integrated supply chain. Subsequently, Lane was senior vice president for integrated supply chain at Motorola for two years before joining Apple in 2008.

At Apple, Lane assumed the role of vice president of operations, overseeing supply chain management and manufacturing for various products, including iPads, iMacs and Apple's Mac accessories. Purdue honored Lane as a Distinguished Engineering Alumna in 2014.

U.S. DEPARTMENT OF DEFENSE'S DIRECTOR OF SPACE POLICY

Travis Langster earned a bachelor's degree in aeronautical and astronautical engineering in 1994. He began his career at General Electric Aerospace, where he worked as an engineer focused on the launch and on-orbit operations of U.S. Air Force satellite systems.

Langster then transitioned to Ansys Government Initiatives (AGI), where he ascended to vice president for DOD and Intel Space. In this capacity, he played a crucial role in leading strategic initiatives and fostering partnerships in the defense and intelligence sectors, contributing to advancement of space technology and national security efforts. Langster later served as vice president and general manager of AGI's COMSPOC, a renowned firm specializing in responsible space behavior and best practices in national security.

In 2022, Langster joined the U.S. Department of Defense as director of space policy, overseeing critical areas including space, nuclear weapons, cyber, missile defense, electromagnetic warfare and countering weapons of mass destruction. Purdue honored Langster with an Outstanding Aerospace Engineer Award in 2021.

ISTOCK/PHOTOFIRES

ADVISOR TO ITALY ON STABILIZING THE LEANING TOWER OF PISA

Gerald A. "Jerry" Leonards earned a master's degree and a PhD in chemical engineering in 1948 and 1952, respectively. As an engineering faculty member at Purdue from 1952 to 1991, he gained recognition as a towering figure in geotechnical engineering. Throughout his career, Leonards' insight and expertise were sought for earthwork and foundation projects worldwide.

He was the only non-European invited by the Italian government to sit on an official commission tasked with investigating methods to stabilize the Leaning Tower of Pisa. His seminal book, *Foundation Engineering*, became a globally recognized reference in the field. In 1980, Leonards was invited by the American Society of Civil Engineers to deliver its prestigious Terzaghi Lecture. His contributions were further recognized with his induction into the National Academy of Engineering in 1988, and receipt of the ASCE's Karl Terzaghi Award in 1989 and Norman Medal in 1965.

Leonards was so beloved and respected that long after he died in 1997 at 75, his 100th birthday was celebrated on April 29, 2021, at an event hosted by the Purdue Geotechnical Society.

ONE OF THE WORLD'S BEST DESIGNERS OF MAJOR BRIDGES

Fritz Leonhardt pursued graduate studies in the College of Engineering from 1932–33, and Purdue awarded him an honorary doctorate in 1980.

An early project during his engineering career in Germany was a collaboration with Wolfhart Andra to design Europe's largest suspension bridge, the Cologne-Rodenkirchen Bridge. This partnership evolved into Leonhardt, Andra and Partners, which became synonymous with designing major bridges worldwide.

Leonhardt's portfolio includes the design of America's first cable-stayed bridge (over the Columbia River in Washington state), the Helgeland Bridge in Norway and several cable-stayed bridges in Dusseldorf, Germany.

He also contributed significantly to Europe's architectural landscape with structures like the Stuttgart Television Tower, Hamburg's Alster-Schwimmhalle and Germany's Olympic Pavilion at the Olympic Games in Montreal (1967) and Munich (1972). The National Academy of Engineering named Leonhardt a foreign associate member in 1983.

CEO OF MULTIPLE BUSINESSES, SMALL AND LARGE

Emily M. Liggett earned a bachelor's degree in chemical engineering in 1977. Her career has been marked by relentless pursuit of excellence and innovation across various industries. Following graduation, Liggett began her career at DuPont Company, where she served as a field engineer, gaining valuable experience in the chemical industry.

In 1980, Liggett joined Raytheon Corporation, where she spent 17 years in dynamic global leadership roles. In 1997, she became CEO of Elo TouchSystems, a pioneering manufacturer of touch screen technology. Under her visionary leadership, the company experienced exponential growth, expanding its global footprint across Asia and Europe while achieving remarkable yearly revenue and profit increases.

Liggett's entrepreneurial spirit and management prowess propelled her to subsequent leadership positions, including managing partner of Tyco Ventures, as well as CEO of Capstone Turbine, Apexon and NovaTorque. Purdue presented Liggett with an Outstanding Chemical Engineer Award in 1999.

ONE OF THE MOST POWERFUL PEOPLE IN THE U.S. WIRELESS INDUSTRY

Roger Linquist graduated with bachelor's (1961) and master's (1963) degrees in mechanical engineering. He was deeply intrigued by cellular communications and shaped an extraordinary career in this field. In 1988, he founded PageMart Wireless and assumed the roles of chief executive officer and, later, chairman. Five years afterward, Linquist co-founded MetroPCS Communications, which he expanded into a company with 24,000 employees.

In 2013, Linquist orchestrated the merger of MetroPCS with T-Mobile, resulting in the sixth-largest wireless company in the world, with more than 100 million customers. Purdue recognized Linquist with a Distinguished Engineering Alumnus Award in 2015.

CREATOR OF THE WORLD'S FIRST ARTIFICIAL KIDNEY

Ben J. Lipps earned a bachelor's degree in chemical engineering in 1962. He furthered his education with a National Institute of Health Fellowship at the Massachusetts Institute of Technology, which propelled him into a career focused on bioengineering. Lipps began his professional journey at Dow Chemical, where he made significant contributions, including creation of the world's first artificial kidney.

In 1988, he became chief executive officer at Fresenius USA, the world's largest fully integrated dialysis products and services company. Lipps retired as chairman and CEO of Fresenius in 2012. Purdue honored him with a Distinguished Engineering Alumnus Award in 2003.

Professor Joseph Liston and aviation technology student Robert Bass

PIONEER IN LANDING ON AN AIRCRAFT CARRIER

Joseph Liston earned bachelor's and master's degrees in mechanical engineering in the 1930s. As a naval aviator, Liston pioneered developing techniques for landing on aircraft carriers—a critical milestone in aviation history. Committed to Purdue beyond his student years, he was a College of Engineering faculty member from 1937 to 1972.

Renowned as "Piston Joe," Liston focused his research on advancing aircraft engine technology, contributing significantly to the field. Moreover, he played a crucial role in establishing Purdue's School of Aeronautics in 1945.

CREATOR OF ONE OF *TIME*'S BEST INVENTIONS OF 2023

Luna Lu, the Indiana ACPA Professor in Concrete Paving and Materials Science, is a pioneer in infrastructure technology. She founded WaveLogix, revolutionizing how engineers monitor and assess concrete structures, particularly on interstates. Lu developed a smart sensor, embedded in poured concrete, that communicates real-time data to engineers via a smartphone app.

This innovation was recognized as one of the Best Inventions of 2023 by *Time* and one of the Next Best Things in Tech for 2022 by *Fast Company*. The American Society of Civil Engineers also named the invention an Infrastructure Gamechanger in 2021. Additionally, Lu's contributions earned her the American Society of Civil Engineers' Alfred Noble Prize in 2022 for a seminal paper detailing her discovery, as well as an Edison Award (considered an "Oscar of innovation") in 2024.

Along with her entrepreneurial work and professorship, Lu is Purdue's vice president for industry partnerships and director of Purdue's Center for Intelligent Infrastructure, leading research and innovation in civil engineering.

CELEBRATING OUR CONSEQUENTIAL IMPACT ON THE WORLD

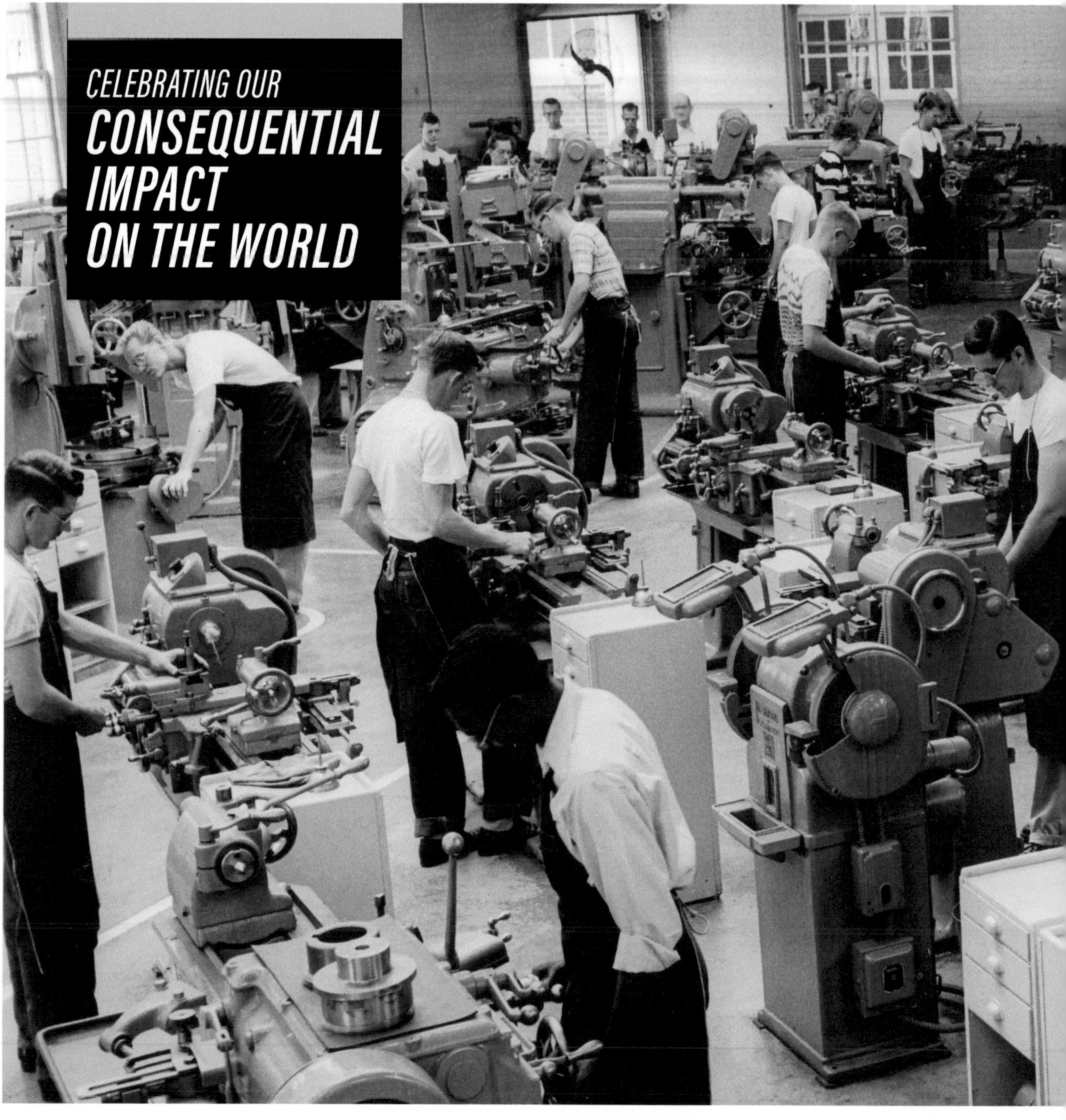

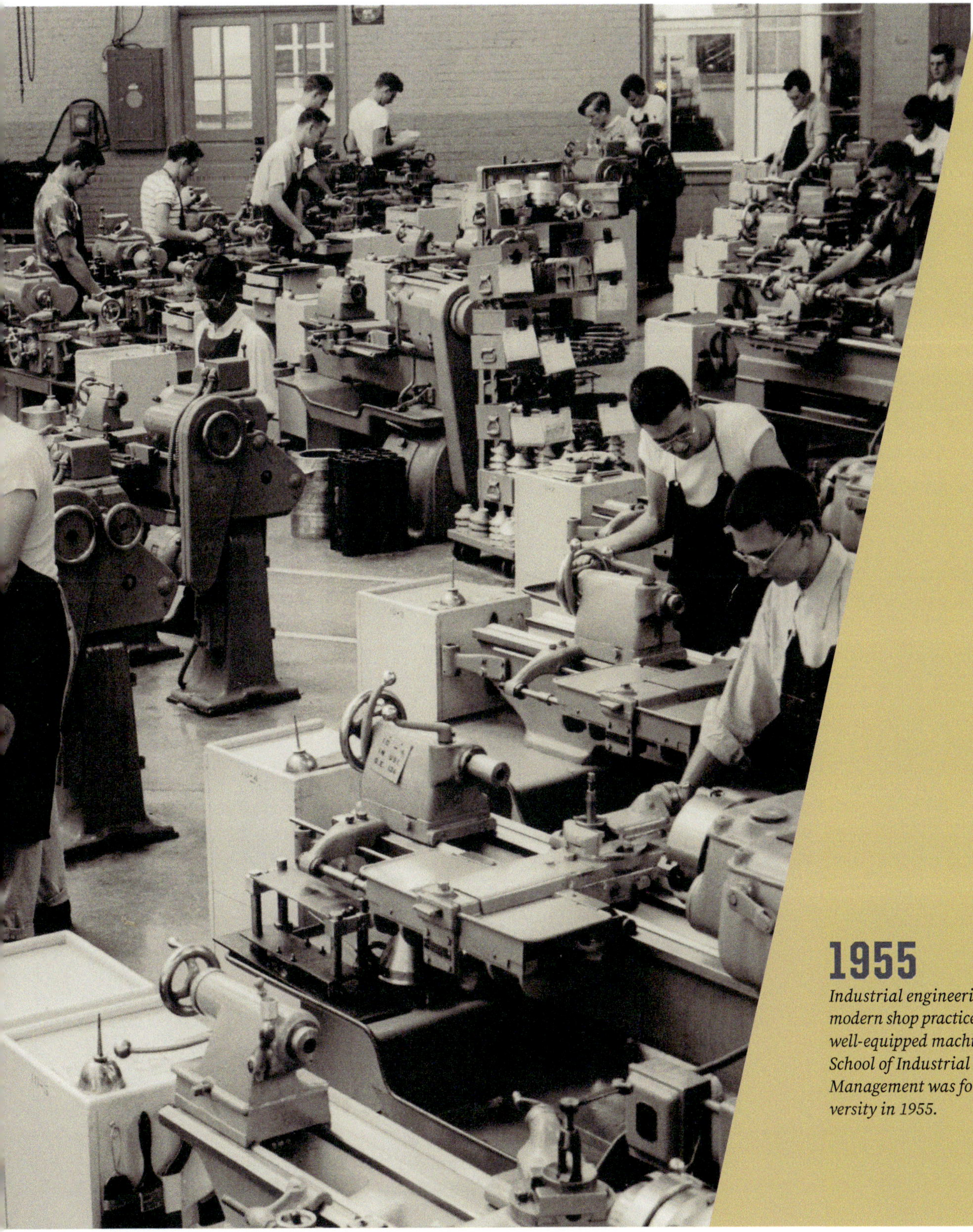

1955

Industrial engineering students learn modern shop practices in one of Purdue's well-equipped machine shops. The School of Industrial Engineering and Management was founded at the university in 1955.

"FATHER OF THE SCIENCE OF SEEING"

Matthew Luckiesh graduated with a bachelor's degree in electrical engineering in 1909. Throughout his career, Luckiesh dedicated his efforts to General Electric. As director of GE's Lighting Research Laboratory from 1924 to 1949, Luckiesh focused his research on "light and vision," causing him to become known as the "Father of the Science of Seeing." He was particularly interested in determining optimal conditions for visibility.

Luckiesh's results included 11 U.S. patents, as well as authorship of 28 books plus more than 800 scientific and technical articles. His designs were highly successful commercially, representing almost 35% of all lamps purchased in America in 1927. In 1929 alone, 13 million lamps designed by Luckiesh were sold. Luckiesh left a legacy of innovation and excellence in lighting and vision science.

TOP EXECUTIVE FOR SOME OF THE WORLD'S TOP BRANDS

Joseph T. Mallof, a 1972 graduate with a bachelor's degree in industrial engineering, has had a distinguished career in the consumer products industry. After earning an MBA from the University of Chicago in 1974, Mallof began a 21-year tenure at Procter & Gamble, where he held senior management roles and contributed to the success of various iconic brands.

In 1995, he joined S.C. Johnson, where he served as president of North American Consumer Products, president of Central and South America, and president of Asia. In 2002, Mallof became president of CIBA Vision. Four years later, he was

appointed chief executive officer of World Kitchens, overseeing such brands as Corelle, Corning-Ware, Magnalite and Chicago Cutlery. Mallof's dedication to education and Purdue is evident through his philanthropic efforts, including the Professor James W. Barany Scholarship endowment to honor his favorite professor.

Purdue recognized Mallof with an Outstanding Industrial Engineering Alumnus Award in 2005.

ACADEMY AWARD WINNER

Bala S. Manian earned a PhD in engineering optics in 1971. He pursued a remarkable entrepreneurial journey in Silicon Valley, where he founded several groundbreaking life science and medical technology companies, including Digital Optics Corporation, Molecular Dynamics, Lumisys and Quantum Dot Corporation.

One of Manian's notable achievements was founding Digital Optics, which developed the first three-color laser film reader/writer system working with Industrial Light and Magic Company (a division of Lucasfilm). This technology played a role in

generating special effects in numerous blockbuster movies, such as *Indiana Jones and the Last Crusade* and *Return of the Jedi*.

Manian's innovative contributions to the film industry were recognized with an Academy of Motion Pictures Arts & Sciences technical award for special effects advances in 1999. Purdue honored Manian with a Distinguished Engineering Alumnus Award in 2000.

SENIOR LEADER OF THE CIA, U.S. AIR FORCE AND NASA

John Joseph Martin earned a PhD in mechanical engineering in 1951. After graduating, he worked for North American Aviation as an engineer for two years before starting a seven-year tenure at Bendix Corporation. Next, Martin transitioned to a role at The Institute for Defense Analyses, focusing on national security issues. During a sabbatical at the Royal Aircraft Establishment in Great Britain, he wrote a book titled *Atmospheric Reentry,* which became widely regarded as one of the best resources on designing reentry vehicles.

In 1969, Martin became a science advisor to President Richard Nixon. Later, in succession, he assumed the roles of associate deputy director of the Central Intelligence Agency, deputy assistant secretary of the U.S. Air Force and associate administrator at NASA.

A DIRECTOR OF 80 TECHNOLOGY COMPANIES

John Martinson obtained a master's degree in aeronautical and astronautical engineering in 1971, following his graduation from the U.S. Air Force Academy the previous year. During his military service, he completed 500 combat missions in Vietnam, earning the Distinguished Flying Cross. Martinson pursued a career in management and marketing at American Cyanamid before gaining entrepreneurial experience at Exxon Enterprises.

In 1979, he joined InnoVen Group, a national venture capital firm based in Saddle Brook, New Jersey, which attracted major investments from such corporations as Monsanto, Emerson Electric and Anheuser-Busch, all based in St. Louis. In 1986, Martinson established his own venture capital firm, Edison Partners, which has since invested more than $2 billion in 230 companies along the East Coast, focusing on software and services for health care, education, finance, marketing and business applications. Additionally, Martinson Ventures, his latest investment firm, serves as a primary investor in 42 emerging companies. He also has invested in 35 private companies and 25 private equity partnerships, as well as served on the boards of 80 companies.

Among accolades, Martinson was recognized as the 18th Distinguished Graduate of the Air Force Academy. Today, Purdue is home to the John Martinson Honors College, ranked among the top 20 honors colleges in the country by College Transitions, and the John Martinson Entrepreneurial Center.

CHAIRMAN AND CEO OF A PUBLIC COMPANY WITH 53,000 EMPLOYEES

Matthew E. "Matt" Massengill received a bachelor's degree in electrical engineering in 1983. He enjoyed a long, successful career at Western Digital Corporation, rising to president, CEO and chairman of the board of this publicly traded company with 53,000 employees.

During his Western Digital tenure, Massengill orchestrated a remarkable turnaround; he is credited with achieving the industry's most efficient cost structure and restoring profitability to the company. In addition, he has served on the boards of GTAT Corp, Vizion Holdings, Conexant Systems, ViewSonic Corporation and Microsemi Corporation. Purdue honored Massengill with an Outstanding Electrical Engineer Award in 1998 and a Distinguished Engineering Alumnus Award in 2002.

CEO OF THE UNITED SPACE ALLIANCE

Michael J. McCulley graduated with bachelor's and master's degrees in metallurgical engineering in 1970. Prior to attending Purdue, McCulley had begun serving on submarines in the U.S. Navy.

After completing his Purdue education, McCulley returned to the Navy and continued his service as a naval officer, aviator and test pilot. Throughout his naval career, he accumulated more than 5,000 flight hours and nearly 400 carrier landings on six aircraft carriers.

In 1984, NASA selected McCulley as a pilot for future Space Shuttle missions. He made his debut in October 1989 as part of mission STS-34, during which the Space Shuttle successfully launched the Galileo spacecraft on its journey to explore Jupiter. In 1996, following his NASA tenure, McCulley joined the United Space Alliance, a company established to manage the Space Shuttle program. He was appointed CEO of the United Space Alliance in 2003.

In addition to his career achievements, McCulley has been honored for his contributions in various capacities. He served as an Olympic torchbearer in the 1996 Atlanta Games, was honored as a Silver Knight of Management from the National Management Association and received a NASA Space Flight Medal.

Purdue recognized him with an Astronaut Engineering Alumnus Award in 1990 and a Distinguished Engineering Alumnus Award in 1998.

INVENTOR OF THE GLOBALLY UBIQUITOUS BARCODE

Paul McEnroe obtained a master's degree in electrical engineering in 1960 while working as an associate engineer at The Boeing Company. Subsequently, he joined IBM as a lab director. At IBM, McEnroe and his team transformed the retail industry by inventing the barcode, now ubiquitous on products worldwide. Additionally, they developed the magnetic code and strip for retail use, as well as barcode scanners, both hand-held (for which McEnroe received the patent) and fixed, integrated into checkout stands.

McEnroe later advanced to IBM's group director of systems development, responsible for overseeing creation of 90% of IBM's products worldwide. Purdue honored McEnroe with a Distinguished Engineering Alumnus Award in 2018.

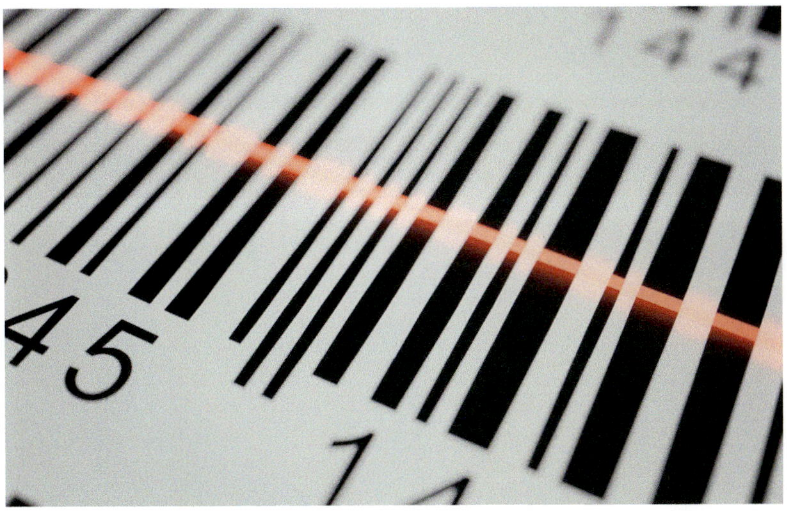

YOUNGEST PURDUE ENGINEERING GRADUATE—AT 16

Christopher Paul "Chris" McNett stands out as a unique graduate of Purdue's College of Engineering. First, he enrolled when he was 13, having earned a perfect SAT score and bypassed Purdue's first-year chemistry requirement at 12.

McNett completed his bachelor's degree in electrical and computer engineering in only three years (receiving it in 2003 at 16)—attaining a remarkable 3.97 GPA and becoming the youngest graduate in the history of the college.

Each summer during his undergraduate years, McNett interned at Microsoft, where he focused on writing security code and developing software for Windows XP. He also earned a patent for layering sound-absorption panels while studying at Purdue.

After graduating, McNett began his professional career as an engineer at Palm, where he specialized in system software and development tools for smartphones. He later earned a master's degree in computer science from Stanford University, followed by a law degree from Harvard University. Currently, McNett is a principal and intellectual property lawyer at McKool Smith in Washington, D.C.

ISTOCK/SEAN PAVONE

CHIEF ENGINEER FOR THE HOOVER DAM

Elwood Mead earned a bachelor's degree in agriculture (1882) and a master's degree in civil engineering (1888). He received an honorary doctorate in 1904.

Mead began his career as a professor at Colorado Agricultural College, teaching the first class in irrigation engineering in the U.S. He then moved into public service, working as Wyoming's state engineer from 1888 to 1899.

Next, he was appointed head of irrigation investigations for the U.S. Department of Agriculture. In 1911, Mead joined the University of California faculty as a professor in rural institutions. His contributions to engineering and public service were recognized further when he was appointed commissioner of the U.S. Bureau of Reclamation in 1924.

As commissioner, Mead served as chief engineer for construction of the Hoover Dam, one of the most iconic engineering feats of the 20th century. Mead's legacy as an engineer, educator and public servant continues to be celebrated. In honor of his significant impact, the reservoir formed by the Hoover Dam on the Colorado River was named Lake Mead.

CO-FOUNDER OF ARUBA NETWORKS —WHICH HE SOLD FOR $3B

Keerti G. Melkote earned a master's degree in electrical engineering in 1993. He then began his career by joining Intel Corporation as a network architect/engineer. After two years, he moved to Cisco Systems Inc., where he served as a product line manager for workgroup switches.

In 1998, Melkote assumed the role of senior director of product management at Shasta Networks. Following his tenure there, he co-founded Aruba Networks in 2002. Five years later, Melkote led Aruba Networks through its initial public offering, and in 2015, he orchestrated the company's acquisition by Hewlett-Packard for $3 billion.

Post-acquisition, Melkote contributed to Hewlett-Packard as president of the Intelligent Edge division until his retirement in 2019. Purdue recognized Melkote with a Distinguished Engineering Alumnus Award in 2017.

PIONEER IN DEVELOPING THE ELECTRONIC ORGAN AND ELECTRONIC PIANO

Benjamin F. Miessner studied electrical engineering at Purdue from 1913 to 1916. In the late 1920s, he demonstrated his entrepreneurial spirit by selling 50 of his patents to the Radio Corporation of America for $750,000. With this capital, he established his own company, Miessner Inventions Inc.

Over the next three decades, Miessner emerged as a pioneer in developing electronic instruments, radio receivers, aircraft radios and phonography, among other innovations. One of his notable patents contributed to the Everett Piano Company's creation of the first electronic organ. In addition, Wurlitzer used a Miessner design to develop its highly successful electronic piano.

Upon winding down his company, Miessner sold nearly 150 additional patents, many related to electronics, sound and music. These patents encompassed inventions including an apparatus for adjusting the tension of tennis racket strings, as well as a nonleaking pen.

PRINCIPAL DIRECTOR OF SPACE

Lindsay Millard, who earned a PhD in aeronautics in 2008, serves as the principal director of space in the Office of the Undersecretary of Defense for Space and Engineering. In this capacity, Millard oversees the Department of Defense's research and engineering initiatives in space, aimed at maintaining the nation's competitive edge in the domain.

Before assuming her current role, Millard was program manager at the Defense Advanced Research Projects Agency (DARPA), where she focused on enabling significant capabilities on small satellites. Before her DOD tenure, Millard was an engineer at the RAND Corporation, where she led studies for Project Air Force and the National Security Research Division, contributing to critical research efforts. Purdue honored Millard with an Outstanding Aerospace Engineer Award in 2019.

GOOGLE VP FOR WORLDWIDE OPERATIONS

Jim Miller, who obtained a bachelor's degree in aeronautical and astronautical engineering in 1986, has had a career spanning four decades at the forefront of high-technology industries. He began at Intel Corporation as a production manager during the introduction of the Pentium chip. Miller then transitioned to Amazon, where he contributed to the remarkable growth of e-commerce.

Following his tenure at Amazon, he joined Cisco Systems Inc., capitalizing on the burgeoning telecommunications industry driven by the internet. He then moved to First Solar, aligning with the increasing demand for green technologies. Subsequently, Miller assumed the role of vice president for worldwide operations at Google, playing a key role in the company's global prominence in search and cloud computing.

In 2018, Miller took the helm as CEO of AREVO, a leading company in 3D printing technology. Two years later, he became chief technology officer of Wayfair, one of the world's largest home-furnishing destinations. Purdue honored Miller with a Distinguished Engineering Alumnus Award in 2014.

LAUNCHER OF 50 COMPANIES

Burton D. Morgan graduated with a bachelor's degree in mechanical engineering in 1938. His professional journey began with positions at the BFGoodrich Company and Pennsylvania Rubber Company before he joined Johnson & Johnson. At J&J, Morgan worked on developing a machine to produce Band-Aids. In 1949, while still at J&J, he ventured into entrepreneurship with the creation of a self-adhesive bra. Although that innovation did not succeed, the setback sparked Morgan's entrepreneurial career.

In 1956, he co-founded Fasson, a company specializing in pressure-sensitive adhesives. This venture evolved into Morgan Adhesives, one of the world's leading manufacturers of such adhesives. Throughout his career, Morgan demonstrated exceptional entrepreneurial acumen, launching 50 companies, six of which grew into major corporations. Purdue awarded Morgan an honorary doctorate in 1992 and named the Burton D. Morgan Center for Entrepreneurship after him.

FIRST WOMAN TO FLY TO SPACE IN A COMMERCIAL VEHICLE

Beth Moses graduated with bachelor's and master's degrees in aeronautical and astronautical engineering in 1992 and 1994, respectively. With her co-op experience included, she spent more than 20 years at NASA, where she served as the Space Station EVA system manager, contributing to the International Space Station (ISS) from its conceptualization to completion.

After the ISS project, Moses transitioned to the private sector, joining Virgin Galactic in 2013 as chief astronaut instructor and interior program manager. On Feb. 22, 2019, she became the 25th Purdue alum to journey into space when she traveled aboard Virgin Galactic's flight VSS Unity 22, reaching nearly 56 miles above Earth.

Notably, she became the first woman to fly to space on a commercial vehicle, earning her Commercial Astronaut Space Wings from the Federal Aviation Administration. Moses has since completed six space flights with Virgin Galactic. Purdue honored her with an Outstanding Aerospace Engineer Award in 2018.

PRIME MINISTER OF JORDAN

Marwan Jamil Muasher earned a bachelor's (1977), a master's (1978) and a PhD (1981) in electrical engineering. He also received an honorary doctorate in 1999. Following his studies, Muasher embarked on a diverse career path. Returning to Jordan, he initially worked at the Jordan Electric Power Company and the Research Institute at King Fahd University of Petroleum and Minerals. He also founded the Jordan Computer Society and wrote political columns for *The Jordan Times,* the sole English-language newspaper in Jordan.

He transitioned to government service, holding such positions as press advisor in the prime minister's office and director of the Jordan Information Bureau in Washington, D.C. In 1995, Muasher made history as Jordan's first ambassador to Israel. During his tenure from 1997 to 2002 as Jordan's ambassador to the U.S., he played a key role in negotiating the Jordan-U.S. Free Trade Agreement.

Upon his return to Jordan in 2002, Muasher was instrumental in shaping the Arab Peace Initiative and the Middle East roadmap. In 2004, he assumed the position of deputy prime minister of Jordan, followed by a brief tenure as prime minister. Subsequently, Muasher served as vice president of external affairs for the World Bank before moving to his current role as vice president for studies at the Carnegie Endowment for International Peace. Purdue presented him with an Outstanding Electrical and Computer Engineer Award in 1999.

ONE OF THE GREATEST ALL-AROUND COLLEGE ATHLETES

Elmer Quillen "Catchy" or "Ollie" Oliphant, a 1916 mechanical engineering graduate, made a substantial impact on Purdue sports. He earned nine varsity letters in football, basketball, baseball and track, becoming Purdue's first athlete to do so. While he also competed in swimming, boxing and wrestling, varsity letters weren't awarded for those sports at the time.

Oliphant's Purdue record for scoring five touchdowns and 43 points in a single game in 1912 still stands. He was Purdue's first All-American, honored in football and basketball. After Purdue, Oliphant attended West Point, earning 11 varsity letters in football, basketball, baseball and track. Despite being 5 feet, 10 inches tall and 175 pounds, he became the heavyweight boxing champion at West Point. He was part of Army's 1914 and 1916 national championship teams and was named to college football's All-Time All-America team for its first 50 years.

Oliphant also set a world record in the 220 hurdles on a grass track. Following graduation from West Point in 1918, Oliphant served in World War I before a successful professional football career, in which he earned All-Pro honors. His contributions to sports brought him induction into the National Football Hall of Fame, College Football Hall of Fame and Purdue Athletics Hall of Fame.

NASA

LEADER OF NASA GALILEO PROJECT'S JUPITER MISSION

William J. "Bill" O'Neil earned a bachelor's degree in aeronautical and astronautical engineering in 1961. Two years later, he commenced his career by joining NASA's Jet Propulsion Laboratory, where he assumed various prominent roles. Notably, O'Neil served as chief of navigation for the Mariner Mars mission in 1971 and for Viking (marking the first U.S. landing on Mars) in 1972.

In 1980, O'Neil took on his most significant mission at NASA, leading the exploration of Jupiter. This endeavor materialized into the Galileo project, which launched on Oct. 18, 1989. The spacecraft, measuring 17 feet long and weighing 2.2 tons, aimed to launch a probe that would orbit Jupiter and analyze its atmosphere. With a $1.3 billion budget, O'Neil served as project manager, overseeing the entire mission.

After a six-year journey covering 2.4 billion miles, the Galileo spacecraft reached Jupiter on July 7, 1995. Originally slated to conclude in 1997, the mission was extended for six years, encompassing three more missions. Ultimately, on Sept. 21, 2003, the Galileo spacecraft was deliberately plunged into Jupiter's atmosphere.

In 1997, O'Neil presented Pope John Paul II with an album featuring the most spectacular images captured by the Galileo spacecraft. That year, Purdue recognized O'Neil with a Distinguished Engineering Alumnus Award.

FIRST WOMAN TO BE CEO OF TWO FORTUNE 500 COMPANIES

Patricia K. "Patti" Poppe graduated with bachelor's and master's degrees in industrial engineering in 1989 and 1991, respectively. She began her career at General Motors, rising to director of operations at the Grand River Assembly Plant. In 2005, Poppe moved to DTE Energy as director of enterprise performance excellence. Six years later, she joined Consumers Energy Company as vice president for customer experience and excellence.

By 2015, Poppe had been promoted to senior vice president, overseeing distribution operations, engineering, and transmissions. The following year, she was appointed president and chief executive officer—titles she held until late 2020 when she became CEO of Pacific Gas & Electric Company, where she remains. Poppe has the distinction of being the first woman to serve as CEO of two Fortune 500 companies. Purdue honored her with a Distinguished Engineering Alumna Award in 2017 and named her an Old Master in 2019.

BLUE ORIGIN'S VP OF MISSION AND FLIGHT OPERATIONS

Audrey Powers earned a bachelor's degree in aeronautical and astronautical engineering in 1999. She started her career at NASA, where she worked on the International Space Station, accumulating more than 2,000 hours in Mission Control. Powers later transitioned to Lockheed Martin, focusing on government satellite programs, while also pursuing a law degree at Santa Clara University.

Upon completing her legal studies, Powers joined Blue Origin in 2013 as vice president of legal and compliance. In this role, she handled regulatory affairs, legislative and policy matters, and supplier and customer negotiations. In June 2021, Powers became vice president of New Shepard mission and flight operations at Blue Origin, overseeing flight operations, vehicle maintenance and launch infrastructure.

On Oct. 13, 2021, she accompanied 90-year-old William Shatner, known as Star Trek's "Captain Kirk," and two others aboard Mission NS-18—an 11-minute trip to space. This made her the 27th Purdue graduate to enter outer space. Purdue presented her with an Outstanding Aerospace Engineer Award in 2021.

BUILDER OF ONE OF THE LARGEST AFRICAN AMERICAN-OWNED COMPANIES IN AMERICA

Mamon M. Powers Jr. earned a bachelor's degree in civil engineering in 1970. After briefly working for Amoco following graduation, he joined his father's company, Powers and Sons Construction, in 1971. Initially serving as the company's treasurer, the younger Powers was appointed president in 1987.

Under his leadership, Powers and Sons Construction changed its focus from residential to commercial and industrial construction, expanding its operations to become one of the largest African American-owned companies in the U.S. Recognizing his contributions to the industry, Indiana Gov. Evan Bayh appointed him to the Purdue Board of Trustees in 1996. Powers served on the board for four additional terms, retiring as vice chair in 2011.

Throughout his career, Powers has contributed his time, skills and resources to various boards and organizations within his community. He continues to serve as CEO and chairman of Powers and Sons Construction. Purdue honored him with a Distinguished Engineering Alumnus Award in 2003 and awarded him an honorary doctorate in 2014.

WINNER OF A NOBEL PRIZE IN PHYSICS

Edward M. Purcell graduated with a bachelor's degree in electrical engineering in 1933. Subsequently, he obtained a master's and a PhD in physics from Harvard University. During World War II, Purcell contributed significantly to the war effort by working at the MIT Radiation Laboratory, where he was involved in developing microwave radar.

In 1946, while conducting research at Harvard with two colleagues, Purcell made a groundbreaking discovery: nuclear magnetic resonance (NMR). This finding laid the foundation for magnetic resonance imaging (MRI), one of the most important medical advancements of the 20th century. For this contribution to science, Purcell and a colleague were awarded the Nobel Prize in Physics in 1952.

Beyond his scientific achievements, Purcell served as a science advisor to Presidents Eisenhower, Kennedy and Johnson, highlighting his influence and expertise in shaping national policy. Purcell received numerous honors, including the Oersted Medal in 1967 and the National Medal of Science in 1972. Purdue awarded him an honorary doctorate in 1970, underscoring his legacy and impact on academia and society.

HOLDER OF 88 U.S. PATENTS—WITH INVENTIONS INCLUDING "REFLECTIVE TRAFFIC EYES"

David E. Ross earned a bachelor's degree in mechanical engineering in 1893. Remaining deeply connected to his hometown of Lafayette, Ross began his entrepreneurial journey in 1905 when he earned three patents for innovative work in automotive gears. Establishing Ross Gear and Tool Company in 1906, he quickly became a key supplier to industries like trucking, racing and yachting.

In 1919, Ross founded Fairfield Manufacturing to specialize in differential gears, while Ross Gear and Tool focused on steering gears, demonstrating his strategic vision and commitment to excellence. With 88 U.S. patents, Ross is known particularly for inventing "reflective traffic eyes."

Beyond his business ventures, Ross played a key role in shaping Purdue's trajectory. His appointment to the Purdue Board of Trustees in 1921 marked the start of transformative initiatives, including development of a new football stadium dedicated to him and trustee George Ade in 1924. Ross' tenure as president of the Board of Trustees brought significant advancements, such as those involving the Purdue Research Foundation, Purdue University Airport and civil engineering camp—all made possible by his generous land donations.

Ross' profound impact on Purdue's growth and development was recognized upon his passing in 1943. He was laid to rest on Slayter Hill, overlooking the campus he cherished, in a poignant tribute to his legacy and commitment to Purdue.

INVENTOR OF THE WORLD'S "WHITEST PAINT" –ONE OF *TIME*'S BEST INVENTIONS OF 2023

Xiulin Ruan, a professor of mechanical engineering, developed what Guinness World Records recognized as the world's "whitest paint." Ruan dedicated six years to this project in his laboratory, aiming to address global warming. If applied to 1,000 square feet of a roof, this white paint is estimated to provide cooling power of 10 kilowatts, surpassing the cooling capacity of many current home air conditioners.

Among accolades, *Time* cited the innovation as one of the Best Inventions of 2023, *Fast Company* hailed it with a 2023 World Changing Ideas Award and *Newsweek* listed it in 2023 as one of Five Tech Inventions That Could Help Save the Planet.

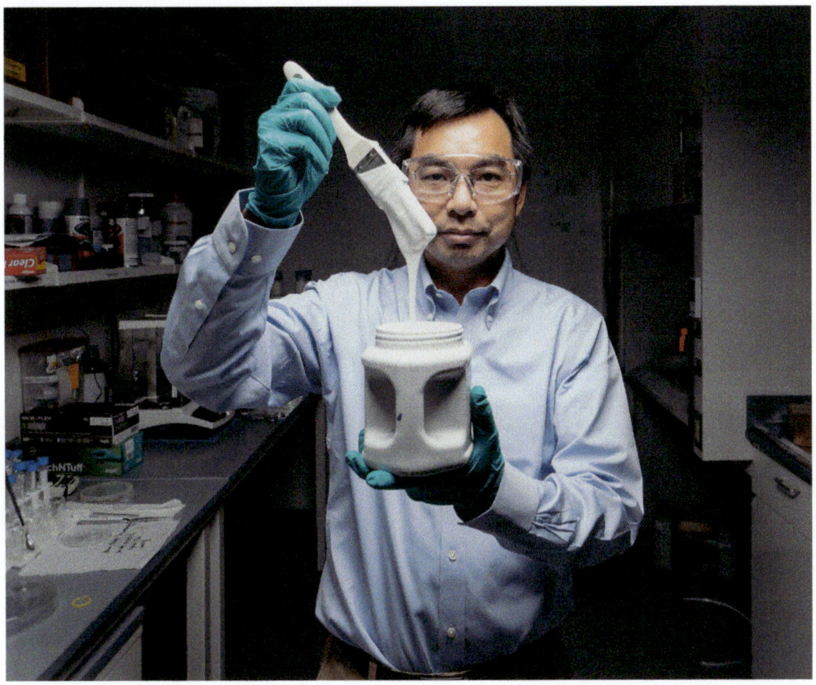

CHAIR AND CEO OF A FORTUNE 500 COMPANY

Jennifer Weerts Rumsey earned a bachelor's degree in mechanical engineering in 1996 and received an honorary doctorate in 2024. She began working as an intern for Cummins Inc. as a Purdue undergraduate. Following in the footsteps of her father, a former Cummins engineer, Rumsey has achieved a series of career milestones at one of the world's largest power companies.

In 2015, Rumsey was appointed chief technical officer, marking the first time a woman held this position in the century-old company. In 2022, she made history again as the first woman to become CEO of Cummins. The following year, Rumsey was the first woman elevated to chair of this publicly traded company with 75,000 employees. She also serves as a board member of Hillenbrand and has been named a fellow of the Society of Automotive Engineers and one of *Fortune*'s 50 Most Powerful Women.

Purdue presented Rumsey with a 40 Under 40 Award, an Outstanding Mechanical Engineer Award in 2016 and a Distinguished Engineering Alumna Award in 2017.

CELEBRATING OUR CONSEQUENTIAL IMPACT ON THE WORLD

1960s

Metallurgical engineering students make assays for gold as part of laboratory work. The School of Metallurgical Engineering was founded on July 1, 1959. It was renamed the School of Materials Engineering in 1973.

INVENTOR OF THE GAMMA-ELECTRIC CELL —AND FIRST AFRICAN AMERICAN PHD IN NUCLEAR ENGINEERING

Henry T. Sampson Jr. graduated from Purdue with a bachelor's degree in chemical engineering in 1956. He began his career as a research chemical engineer at the U.S. Naval Weapons Center China Lake, California, where he contributed to solid-rocket motor research. In 1961, Sampson embarked on graduate studies, receiving two master's degrees, and became the first African American to earn a PhD in nuclear engineering in the U.S. (from the University of Illinois Urbana-Champaign in 1967).

Following his doctoral studies, Sampson joined The Aerospace Corporation as a project engineer, launching a long, impressive tenure at the company. In 1971, he obtained a patent for his invention of a gamma-electric cell, a groundbreaking device that converted gamma ray energy into electricity.

Throughout his career, Sampson continued to innovate, earning additional patents for rocket propellants that transformed the manufacturing of solid-propellant rocket motors. In 1981, Sampson was promoted to the directorate of space test programming at Aerospace, where he served as director of planning and operations and played a pivotal role in leading the senior engineering staff through all phases of satellite planning, launching and space operations. Purdue honored Sampson with an Outstanding Chemical Engineer Award in 2009 and a Distinguished Engineering Alumnus Award in 2013.

LEADER OF BOEING'S 787-10 DREAMLINER CREATION

Kenneth Sanger graduated with bachelor's (1981) and master's (1983) degrees in aeronautical and astronautical engineering. His 35-year career with The Boeing Company culminated in his role overseeing the design, development, production and certification management of the Boeing 787-10 Dreamliner.

As vice president and general manager of 787 Airplane Development, Sanger headed a global team of internal staff and suppliers across North America, Asia, Europe, Russia and the Middle East for this multi-billion-dollar project. The first 787-10 Dreamliner was delivered to Singapore Airlines on March 25, 2018. As of Dec. 31, 2023, there have been orders for 266 Boeing 787-10 airplanes, with 96 delivered. Purdue honored Sanger with a Distinguished Engineering Alumnus Award in 2019.

GLOBAL PIONEER IN DIGITAL PROTECTION

Edmund O. "Ed" Schweitzer III graduated with bachelor's and master's degrees in electrical engineering in 1968 and 1971, respectively. His career began in 1968 at the U.S. Department of Defense, where he spent five years. After obtaining his PhD from Washington State University, Schweitzer taught at Ohio University and later Washington State University, serving as an associate professor at the latter institution until 1985.

In 1982, while at Washington State, Schweitzer invented the SEL-21, the first microprocessor-based protective relay, which transformed the industry. Inspired by this breakthrough, Schweitzer founded Schweitzer Engineering Laboratories (SEL) in his basement in 1984. Today, the company has more than 6,000 employees, with 61 offices in the U.S. and 46 offices globally. In 2019, SEL established a manufacturing plant in the Discovery Park District at Purdue as its fourth manufacturing center. Schweitzer is a fellow of the Institute of Electrical and Electronics Engineers and was awarded its highest honor, the Medal in Power Engineering. In 2022, he was elected a member of the National Academy of Engineering. Purdue presented Schweitzer with a Distinguished Engineering Alumnus Award in 2020.

NATIONAL PARK SERVICE

SUPPLIER OF BLACK GRANITE FOR THE VIETNAM VETERANS MEMORIAL

Manahar R. "Manu" Shah earned a master's degree in mechanical engineering in 1968. Following graduation, Shah joined Ohio Brass as a senior engineer. In 1973, he moved to International Trucks, where he served as a senior project engineer until 1982. During this time, Shah relocated his family from India to the U.S.

In 1981, while running a family business from his home's basement, he successfully bid for the contract to supply black granite for the Vietnam Veterans Memorial. Despite risking his life savings on this project, it proved to be a turning point, leading to establishment of MS International Inc.

The pioneering company has since flourished into a global leader in flooring, wall and countertop surfaces, with more than $2.5 billion in revenue and over 3,000 U.S. employees. Ernst & Young honored Shah with an Entrepreneur of the Year Award in 2007. Purdue presented him with an Outstanding Mechanical Engineer Award in 2018 and a Distinguished Engineering Alumnus Award in 2021. Today, the Shah Family Global Innovation Lab in the Purdue College of Engineering bears his name.

50TH PRIME MINISTER OF EGYPT

Essam Abdel-Aziz Sharaf earned his master's degree (1980) and PhD (1984) in civil engineering. After completing his studies, Sharaf was a visiting professor at Purdue in 1984 before returning to Egypt. He then served as an assistant professor of highway and traffic engineering at Cairo University, followed by a brief stint at King Saud University in Saudi Arabia.

In 1996, Sharaf became a professor of highway engineering at Cairo University and began advising the Egyptian government on transportation matters. In 2004, he was appointed Egypt's minister of transport, a position he held until December 2005, when he returned to Cairo University. As minister of transport, Sharaf undertook unprecedented initiatives to modernize Egypt's transportation infrastructure system, and he became a major force in the country's transportation education, research and public policy.

Egypt's governing military council tasked him to form a government during the 2011 revolution, and Sharaf served as Egypt's 50th prime minister for most of 2011. Purdue honored him with a Distinguished Engineering Alumnus Award in 2006.

ASSISTANT SECRETARY OF DEFENSE FOR ACQUISITIONS

Tanya M. Skeen graduated with a bachelor's degree in aeronautics and astronautics in 1990. Subsequently, she enlisted in the U.S. Navy and was assigned to the Nuclear Power Training Command. After serving in the Navy, Skeen pursued engineering roles in the private sector, contributing to development of such projects as the USAF B-2 bomber, as well as the X-35 concept demonstrator, a precursor to the F-35 Joint Strike Fighter.

Her career later transitioned to the Nevada Test and Training Range, where she was a flight test engineer. In this position, she was responsible for test planning, execution and data analysis for various aircraft, including the B-2, F-117, F-15, F-16 and F-22. In 2004, Skeen joined the Air Force as chief engineer in the USA Rapid Capabilities Office, overseeing a significant portfolio of projects. In 2017, she served as director and project executive officer of the Army's Rapid Capabilities Office. The following year, Skeen joined the U.S. Department of Defense as executive director of the F-35 Lightning II Joint Program Office.

In 2022, she was promoted to assistant secretary of defense for acquisitions, leading a large team of program managers, project officers and engineers. In 2023, Skeen joined CACI International as senior vice president of corporate strategy and development. Among government honors, Skeen has received a David Packard Excellence in Acquisition Award from the DOD, a Joint Civilian Service Achievement Award and an Assistant Secretary of the Air Force Annual Award for Acquisition Process Improvement. Purdue recognized her with an Outstanding Aerospace Engineer Award in 2018 and a Distinguished Engineering Alumna Award in 2021.

INVENTOR OF FIBERGLASS

R. Games Slayter, who graduated with a bachelor's degree in chemical engineering in 1921, is credited with developing the coarse fibers that facilitated commercial production of the first fiberglass product (Fiberglas). Slayter, who played the tuba in Purdue's "All-American" Marching Band, spent his entire career at Owens Corning, where he earned more than 90 patents in fiberglass technology.

Slayter's significant contributions led to his induction into the National Inventors Hall of Fame. At Purdue, a donation from Slayter facilitated construction of the Slayter Center of Performing Arts, completed in 1964.

The Slayter Center of Performing Arts is an outdoor concert bandshell completed in 1964 and dedicated on May 1, 1965. The building was designed to look like Stonehenge in Wiltshire, England.

JULIAN HERZOG

CO-FOUNDER OF A COMPANY HE SOLD FOR $2.1B

David A. Stamm earned a bachelor's degree in electrical engineering in 1973. He was among the initial college recruits to join Intel Corporation, then a newly established company. At Intel, Stamm was instrumental in creating the concept and circuit design for the Intel 8748, one of the most widely "second-sourced" microprocessors in history.

In 1980, he left Intel to co-found Daisy Systems, a pioneering company in computer-aided engineering. Under his leadership, Daisy Systems experienced rapid growth, employing more than 1,000 people by 1985. Following his tenure at Daisy Systems, Stamm co-founded Clarify in 1991. Clarify specialized in providing software solutions to enhance the operations of sales and customer support functions, positioning itself as one of the earliest customer relationship management companies globally.

As chief executive officer and chairman of the board, Stamm steered Clarify to significant growth, achieving a 22% market share in its sector. In 1995, Clarify went public, and in 2002, it was acquired by Nortel for $2.1 billion. Purdue presented Stamm with an Outstanding Electrical and Computer Engineer Award in 1996 and a Distinguished Engineering Alumnus Award in 1999.

PURDUE'S FIRST WOMAN TO EARN AN ENGINEERING DEGREE—IN 1897

Martha Dick Stevens received a bachelor's degree in civil engineering in 1897, becoming the first woman to earn an engineering degree from Purdue. She also earned two other Purdue degrees: a bachelor's degree in mathematics and biology in 1894 and a master's degree in science in 1898.

While a Purdue student, Stevens was president of the Photographic Club, vice president of the Philalethean Literary Society and YWCA, and alumni editor of the *Purdue Exponent*.

After graduating, Stevens taught college mathematics before retiring to care for her aging parents. Purdue's Lyles School of Civil Engineering observed the 120th anniversary of Stevens' graduation with a yearlong celebration in 2016–2017.

"FATHER OF THE 737"

Richard W. "Dick" Taylor earned a bachelor's degree in mechanical engineering in 1942. As a Reserve Officers' Training Corps (ROTC) student, Taylor went directly into service in the Army during World War II, seeing combat during the Battle of the Bulge. After the war, in 1946, he was hired by The Boeing Company, where he became an aviation legend over the next 50 years.

Taylor served Boeing in various roles, including airplane designer; test pilot (including for the B-47 Stratojet, the first large, swept-wing jet-powered plane); aerospace executive; and consultant. He played a pivotal role in developing the 737 aircraft as Boeing's director of engineering. Additionally, he led the certification of a two-person flight crew and was instrumental in development of extended-range twin-engine operations (ETOPS).

Taylor retired in 1991 as a Boeing vice president but was an active pilot for 70 years, making his final flight in December 2014 at 93. Purdue recognized Taylor with an honorary doctorate in 1973 and an Outstanding Aerospace Engineer Award in 1999.

WALMART'S EVP AND CIO

Karenann "Kat" Terrell earned her master's degree in electrical engineering in 1988. Following graduation, she commenced her career at General Motors as an engineer and held various roles for 12 years.

In 2000, Terrell transitioned to Chrysler as director of e-business. Over the years, she ascended the ranks to chief information officer for Chrysler Group and Mercedes-Benz North America. In 2006, Terrell joined Baxter International as corporate vice president and CIO. She then moved to Walmart in 2010, serving as executive vice president for information systems before being promoted to EVP and CIO in 2012.

During her tenure at Walmart, Terrell led a global team of more than 5,000 technologists. From 2017 to 2021, she served as GlaxoSmithKline's chief digital and technology officer. Purdue recognized Terrell with a Distinguished Engineering Alumna Award in 2015.

Brothers Frank Thomas Jr. and Don Thomas

INVENTOR OF THE SOFT-SERVE ICE CREAM MACHINE

Frank Thomas Jr., a 1941 mechanical engineering graduate, is credited with inventing the first patented soft-serve ice cream machine in 1938. He was able to perfect the freezing and agitating process with his machine and produce the popular dessert.

He also played a role in developing a grill to flame-broil beef patties—that prototype grill was made for James W. McLamore and Dave Edgerton, the founders of Burger King.

Thomas distributed his soft-serve ice cream machines and flame-broiling equipment through his firm, the General Equipment Company. In 1957, he opened a demonstration store for his company's products, running the location alongside his brother and fellow Purdue alum Donald Thomas, a 1951 mechanical engineering graduate.

The store's success spurred the creation of the brothers' restaurant chain, Burger Chef. The fast-food franchise operated more than 1,000 locations at its peak and was sold to General Foods, then eventually owned by Hardee's.

CEO OF THE LARGEST FAST-FOOD CHAIN IN THE WORLD

Don Thompson graduated in 1984 with a bachelor's degree in electrical engineering. In 1990, he embarked on a 25-year career with McDonald's Corporation, beginning as a manager of a McDonald's restaurant on the South Side of Chicago. He steadily climbed the corporate ladder, serving as senior vice president of the Midwest region, chief operating officer of all U.S. stores, and eventually COO of the McDonald's global operation.

In 2012, Thompson was promoted to president and CEO of McDonald's Corporation, overseeing its more than 40,000 stores worldwide. After retiring from McDonald's, Thompson established a venture capital firm with his wife, Liz Thompson, who earned a Purdue electrical engineering degree in 1985.

Thompson served as a Purdue trustee from 2009–22. In 2022, the Board of Trustees approved naming the Minority Engineering Program within the College of Engineering for Don and Liz Thompson. Both were active in the Minority Engineering Program as students.

PURDUE'S FIRST AVIATOR
AND AN ADVISOR TO THE WRIGHT BROTHERS

J. Clifford "Turp" Turpin earned a bachelor's degree in mechanical engineering in 1908. After graduating, he returned to his hometown of Dayton, Ohio, where he secured a job with New Era Auto-Cycle, a company specializing in engine-powered bicycles for America's growing motorcycle market. Turpin's connection to Wilbur and Orville Wright led him to assist them with airplane engine issues.

When the Wright brothers established the Wright Company in 1909, they invited Turpin to join them. He was tasked with working on the engines for their planes, and he became a student at the Wright Flying School, learning from Orville. Turpin subsequently became the first Purdue alum to pilot an aircraft and signed on as an instructor at the Wright Flying School.

In 1911, the Wright brothers formed the Wright Exhibition Team, which participated in airshows nationwide. As a pilot on the team, Turpin set a record for flying a plane at 9,400 feet. However, tragedy struck in 1912 when Turpin lost his flying partner and fellow team member in a plane crash at a show in Yakima, Washington. Once hailed as the "King of the Air," Turpin decided not to fly again. He briefly sold cars in New York before relocating to Boston to start a successful coffin business.

DIRECTOR OF GLOBAL VEHICLE ENGINEERING AT FORD MOTOR COMPANY

Donald Allen "Don" Ufford earned a master's degree in mechanical engineering in 1989. He began his 31-year career at the Ford Motor Company as a research engineer in Advanced Body Systems. Later, Ufford progressed to such roles as chief engineer for the Truck Division, chief engineer for North America and chief engineer for Europe.

Eventually, he became director of global vehicle engineering, overseeing Ford engineers and vehicles worldwide. During his time at Ford, Ufford led teams that developed many of the company's top-selling vehicles, which earned numerous Car and Truck of the Year awards. In 2020, he retired from Ford and joined the National Institute of Standards and Technology as assistant director for supply chain and ecosystem management. Purdue presented Ufford with a Distinguished Engineering Alumnus Award in 2017.

GROUP PRESIDENT OF KIMBERLY-CLARK NORTH AMERICA

Kimberly Underhill graduated with a bachelor's degree in chemical engineering in 1987. Afterward, she began her career as a process engineer at General Electric. A year later, Underhill transitioned to Kimberly-Clark Professional, where she joined the brand management team. Over the next 33 years, Underhill steadily progressed through various leadership roles within Kimberly-Clark Professional, culminating in her appointment as global president in 2014.

Underhill's continued success led to her promotion in 2018 to group president of Kimberly-Clark North America, with full responsibility for the division's $8 billion consumer products portfolio. In 2021, Underhill joined the Boston Consulting Group, showcasing her versatility and expertise in the field. Purdue recognized Underhill with a Distinguished Engineering Alumna Award in 2018.

LEADER OF THE WORLD'S SECOND-LARGEST MANUFACTURER OF MICRO MOTORS

Patrick S. Wang earned bachelor's and master's degrees in electrical engineering in 1972. Following his education, Wang returned to his family's business in Hong Kong. With Wang as chairman and chief executive officer, the company, Johnson Electric, transitioned from producing motors for high-end model racing cars to manufacturing motors for hand-held hair dryers—a technology Wang developed.

Through his strategic vision and innovative approach, Johnson Electric has grown exponentially, becoming the second-largest manufacturer of micro motors globally, with nearly $4 billion in annual sales. The company operates manufacturing facilities in 22 countries and employs more than 35,000 people.

Purdue honored Wang with an Outstanding Electrical Engineer Award in 1998 and a Distinguished Engineering Alumnus Award in 2001. Additionally, Purdue's Seng-Liang Wang Hall of Electrical and Computer Engineering is named for Patrick Wang's father.

ONE OF PURDUE ENGINEERING'S PIONEERING WOMEN ELECTRICAL ENGINEERING GRADUATES

Josephine Webb graduated from Purdue's College of Engineering in 1940 with a bachelor's in electrical engineering, becoming one of the early pioneering women to earn a degree in her field from the institution. Webb was one of only five women in engineering at Purdue at that time. She later joined Westinghouse Electric Company as a design engineer, contributing to such projects as the electrical grids of the Coulee, Hoover and Boulder dams.

While working at Westinghouse, she obtained two patents for oil circuit breaker contact design, or "switchgear." Webb's career also included roles at the facsimile development laboratory of the Aiden Products company, where she designed an 18-inch fax machine known for its exceptional resolution. Alongside her husband, Herbert J. Webb—who earned his bachelor's in electrical engineering in 1940, she established a successful consulting firm with clients including The Boeing Company and the U.S. Bureau of Mines.

In 2023, Purdue posthumously honored her as an Outstanding Electrical and Computer Engineer. The Society of Women Engineers, which named Webb a fellow, celebrates her as one of the earliest female electrical engineers in the U.S. She died in 2017 at 98.

DIRECTOR OF ENGINEERING AT NASA'S JOHNSON SPACE CENTER

Julie Kramer White graduated with a bachelor's degree in aeronautical and astronautical engineering in 1990. Her career at NASA began through a co-op experience as a Purdue student. Following graduation, she became a full-time employee at NASA, where she has remained for more than 30 years. Kramer White has made significant contributions to NASA's Human Spaceflight Programs, specializing in structural design, analysis, certification and vehicle systems engineering. She has played important roles in such projects as the Space Shuttle Orbiter, X-38 and the International Space Station.

Additionally, Kramer White was a founding member of NASA's Engineering and Safety Center following the Columbia accident. From 2006 to 2017, she served as chief engineer for the Orion Multi-Purpose Crew Vehicle. Currently director of engineering for the Johnson Space Center, she oversees critical engineering support for all of NASA's human spaceflight programs, including the ISS, Orion, Commercial Crew Program and NASA's latest lunar exploration campaign.

Kramer White has received numerous accolades, including the NASA Space Flight Awareness Silver Snoopy Award. Purdue named her an Old Master in 2015 and recognized her with an Outstanding Aerospace Engineer Award in 2017 and a Distinguished Engineering Alumna Award in 2023.

DEVELOPER OF BRIGHT-RED LED BRAKE LIGHTS

Jerry M. Woodall served as a microelectronics professor at Purdue from 1993 to 1999 and from 2005 to 2012. This prolific inventor and scientist is renowned for his groundbreaking invention of the first commercially viable heterojunction material for bright-red LED lights, which enabled technologies including automobile brake lights, traffic lights, CD and DVD players, TV remote controls and computer networks.

Woodall holds an impressive portfolio of more than 80 patents. He has received numerous accolades, including recognition from IBM and NASA, as well as a National Medal of Technology and Innovation awarded in 2001.

WORLDWIDE ENGINEERING AND MANUFACTURING LEADER AT JOHN DEERE

Adel A. Zakaria earned a master's degree and a PhD in industrial engineering in 1970 and 1973, respectively. After starting his career teaching engineering at the State University College of New York, he joined John Deere in 1976 as a staff engineer, focusing on developing software for machine parts, machine-tool utilization and computer-aided process planning.

In 1995, Zakaria was promoted to senior vice president of engineering and manufacturing for the Worldwide Agricultural Equipment Division at John Deere. Zakaria's leadership resulted in the successful launch of 250 new products, contributing to a doubling of sales to $7 billion annually.

In 1997, he was instrumental in achieving a historic labor settlement between John Deere and the United Auto Workers. Purdue recognized him with an Outstanding Industrial Engineer Award in 1998 and a Distinguished Engineering Alumnus Award in 1999.

ENGINEERING PIONEER AT PURDUE AND HARVARD

Maurice J. Zucrow made history as the first student to earn a PhD from Purdue's College of Engineering, completing his dissertation in mechanical engineering in 1928. Six years earlier, he had achieved another milestone when he became the first to graduate with a bachelor's degree in engineering from Harvard University.

Following his time at Purdue, Zucrow devoted 17 years to industry, contributing to such projects as the nation's inaugural gas turbine engine research and development at Elliott Company, as well as assisting in development of AeroJet's JATO rocket. In 1946, Zucrow moved to academia, joining the faculty of Purdue's School of Aeronautics and Astronautics, where he stayed until his retirement in 1966.

In 1948, he authored *Principles of Jet Propulsion and Gas Turbines*, the pioneering textbook in this field. Concurrently, Purdue established its first rocket propulsion laboratory to support Zucrow's research. Subsequently, the combustion laboratory was erected in 1952 to accommodate Zucrow, his team and their graduate students. The gas turbine laboratory followed in 1954.

In 1965, the high-pressure rocket research laboratory was added, culminating Zucrow's career as he retired a year later. Today, the renamed Maurice J. Zucrow Laboratories is the largest academic propulsion lab in the world. Zucrow's recognition included election to the board of directors of the American Rocket Society in 1957, a Distinguished Civilian Service Award from the Department of the Army in 1967 and an honorary doctorate from Purdue in 1970.

SCHENECTADY

CONSEQUENTIAL LEADERS

The Deans Who Have Guided Purdue Engineering for 150 Years

150 YEARS OF PURDUE ENGINEERING

THE DEANS

OF PURDUE ENGINEERING

Over its 150-year history, Purdue engineering has been led by 13 distinguished individuals, each entrusted with guiding one of the world's premier engineering schools.

As deans, they have overseen all educational, budgetary and administrative affairs of the college. They have also provided the vision and leadership necessary for the Purude engineering to excel and to advance the university's missions in teaching, research and service.

W.F.M. GOSS
1900–1907

CHARLES H. BENJAMIN
1907–1920

HENRY T. YANG
1984–1994

RICHARD J. SCHWARTZ
1995–2001

LINDA P.B. KATEHI
2002–2006

LEAH H. JAMIESON
2006–2017

AUDREY A. POTTER
1920–1953

GEORGE A. HAWKINS
1953–1967

RICHARD J. GROSH
1967–1971

JOHN C. HANCOCK
1972–1983

MUNG CHIANG
2017–2022

MARK S. LUNDSTROM
2020, 2022–2023

ARVIND RAMAN
2023–Present

CONSEQUENTIAL INDIVIDUALS

Purdue Engineering Alumni and Faculty Elected as National Academy of Engineering Members

NATIONAL ACADEMY OF ENGINEERING

PURDUE ENGINEERING ALUMNI AND FACULTY INDUCTEES

Established in 1960, the National Academy of Engineering is home to the world's most accomplished engineers. Election to the NAE is considered to be among the highest honors in engineering-related fields, and it often comes as a recognition of a lifetime's worth of accomplishments.

Nomination for membership can only be done by a current member of the NAE for outstanding engineers with identifiable contributions or accomplishments in one or both of the following categories:

- Engineering research, practice or education, including, where appropriate, significant contributions to the engineering literature
- Pioneering of new and developing fields of technology, making major advancements in traditional fields of engineering or developing/implementing innovative approaches to engineering education

Five Purdue College of Engineering alumni inducted into the National Academy of Engineering also have served as faculty members: Michael Ladisch, David Landgrebe, Mark Lundstrom, Harold Michael, and Raymond Viskanta.

PURDUE ENGINEERING ALUMNI

KRISTI ANSETH

NEIL ARMSTRONG

GEORGE BANKOFF

CHRISTOPHER BURKE

PATRICK CHAPMAN

RAMALINGAM CHELLAPPA

GERALD HINES

JOHN HUDSON

KENNETH IVERSON

JÓN ATLI BENEDIKTSSON

PAUL BEVILAQUA

ANDREW BOBECK

ARINDAM BOSE

CHRISTOPHER BOWMAN

LISA BRANNON-PEPPAS

JENNIFER SINCLAIR CURTIS

MICHAEL ESKEW

WILLIAM GERSTENMAIER

RICHARD GROSH

DEBORAH GRUBBE

ARTHUR HANSEN

CHARLES JAKOWATZ JR.

JOHN KLIER

RICHARD KORSMEYER

MICHAEL LADISCH

DAVID LANDGREBE

FRITZ LEONHARDT

JEFFERSON LIEVENSE

MARK LUNDSTROM

JOHN MCDONALD

HAROLD MICHAEL

STEPHEN MORSE

GEORGE MUELLER

DOUGLAS OWEN

LLOYD ROBESON

ANIL SACHDEV

ANDREW SAGE

THOMAS SHERIDAN

KAY STANNEY

HEMANT THAPAR

RAYMOND VISKANTA

PURDUE ENGINEERING FACULTY

RAKESH AGRAWAL

JAN ALLEBACH

SANTOKH BADESHA

ARDEN BEMENT JR.

STEPHAN BILLER

JOHN BOGDANOFF

WENG CHEW

DALE COMPTON

SUPRIYO DATTA

KING-SUN FU

LESLIE GEDDES

MILTON HARR

KATHLEEN HOWELL

ENRIQUE IGLESIA

LEAH JAMIESON

SANGTAE KIM

MICHAEL LADISCH

DAVID LANDGREBE

GERALD LEONARDS

MARK LUNDSTROM

AJAY MALSHE

HAROLD MICHAEL

BYRON PIPES

LAURA PYRAK-NOLTE

DORAISWAMI RAMKRISHNA

GINTARAS REKLAITIS

GAVRIEL SALVENDY

REINHARDT SCHUHMANN JR.

VLADIMIR SHALAEV

JEFFREY SIIROLA

KUMARES SINHA

JAMES SOLBERG

METE SÖZEN

JOHN SUTHERLAND

RAYMOND VISKANTA

ANDREW WEINER

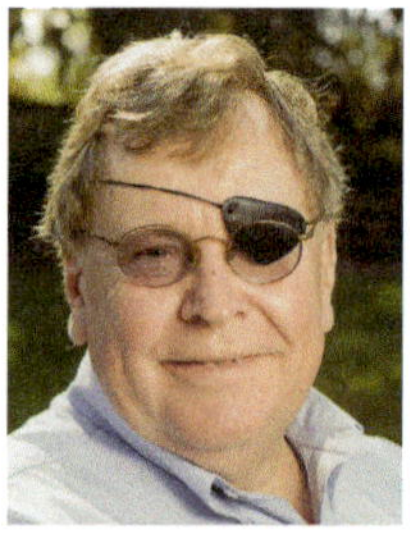
JERRY WOODALL

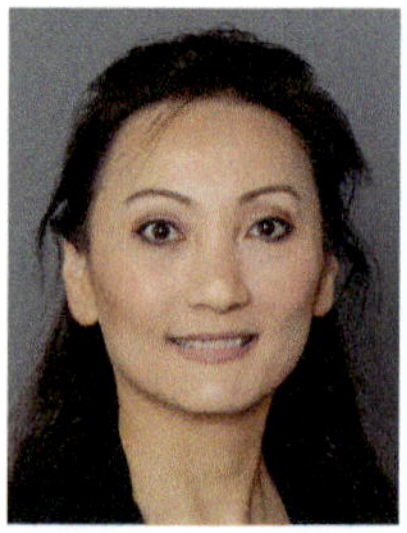
YUEHWERN YIH

1946

Following World War II, returning soldiers studied industrial and mechanical engineering in Purdue's Michael Golden Laboratories.

INDEX

A

Abreu, Sue Hudson, 2
Agrawal, Rakesh, 91
Allebach, Jan, 91
Anderson, Carrie (Byrnes), 2
Anseth, Kristi, 88
Armstrong, Charles G. "Chuck," 3
Armstrong, Neil, 3, 88
Atalla, "John" Mohammed M., 4

B

Babbitt, John F. "Jack," Jr., 4
Badesha, Santokh, 5, 91
Banbury, Fernley H., 5
Bandla, Sirisha, 6
Bankoff, George, 88
Banner, Donald W., 6
Baqai, Farhan A., 7
Bateman, Robert E. "Bob," 7
Beal, Douglas "Woody," 8
Bechtel, Stephen, Jr., 8
Bement, Arden, Jr. 91
Benediktsson, Jón Atli, 89
Benjamin, Charles H., 78
Berlin, Donovan R., 9
Beutner, Thomas, 9
Bevilaqua, Paul, 89
Bhogal, Nikhil, 9
Biller, Stephan, 91
Birck, Michael J., 10
Blalock, Marion Williamson, 10
Bobeck, Andrew, 89
Bock, Charles C. "Charlie," Jr., 11
Bogdanoff, John, 91
Bose, Arindam, 89
Bowen, Robert L. "Bob," 11
Bowman, Christopher, 89
Brannon-Peppas, Lisa, 89
Bridwell, G. Porter, 12
Brown, Mark N., 12
Burgess, Mark A., 13
Burke, Christopher B., 13, 88
Burr, Stacey, 16

C

Caldwell, Orestes Hampton, 16
Cernan, Eugene A. "Gene," 16
Champaigne, Jack, 17
Chapman, Patrick, 88
Chellappa, Ramalingam, 88
Chew, Weng, 91
Chiang, Mung, 18, 85
Chiminski, John R., 18
Chua, Leon O., 19
Compton, Dale, 91
Cooper, Fred, 19
Cooper, James A. "Jim," Jr., 20
Costas, John Peter, 20
Couch, Lana M., 21
Cunningham, Ward, 21
Curtis, Jennifer Sinclair, 89

D

Dankanich, John, 21
Datta, Supriyo, 22, 91
Davis, Darryl W., 22
Diachun, Benjamin L. "Ben," 23

E

Earhart, Amelia, 23
Ebersole, C. Douglas, 24
Edwardson, John A., 24
Ellis, Charles Alton, 25
Elmore, William Bruce "Bill," 25
Endsley, Mica R., 26
Engibous, Thomas J., 26
Esfarjani, Keyvan, 26
Eskew, Michael L. "Mike," 27, 89
Ewry, Ray, 27

F

Fessenden, Reginald, 30
Field, J. Douglas "Doug," 30
Fisher, Jeffrey "Jeff," 31
Forney, Robert C. "Bob," 31
Fu, King-Sun, 91

G

Galloway, Patricia D., 32
Gambaro, Ernest U., 32
Gambone, Mary Lee, 33
Gatewood, Roger B., 33
Geddes, Leslie A., 34, 91
George, Roscoe H., 34
Gerstenmaier, William H. "Bill," 35, 89
Gibbons, James D. "Jim," 35
Gibson, Karen, 36
Gilbreth, Lillian, 36
Goss, W.F.M., 84
Greene, Frank S., Jr. 37
Greenleaf, James F., 37
Grimm, Terry, 38
Grissom, Virgil "Gus," 38
Grosh, Richard J., 85, 89
Grubbe, Deborah, 89

H

Hagle, Marc Lee, 39
Hale, N. Wayne, Jr. 39
Hamid, Mamoon, 40
Hancock, John C., 85
Hansell, Clarence Weston, 40
Hansen, Arthur, 89
Harris, Anthony "Tony," 40
Harr, Milton, 91
Hawkins, George A., 85
Held, Gerald D. "Jerry," 41
Hess, Amy S., 41
Hesse, Walter J., 44
Hines, Gerald, 88
Howell, Kathleen, 44, 91
Hruby, Jill M., 45
Hudson, John, 88

I

Iglesia, Enrique, 91
Iverson, F. Kenneth "Ken," 45, 88

J
Jakowatz, Charles V. "Jack," Jr., 46, 89
Jamieson, Leah H., 84, 91
Jamison, Alpha P., 46
Jazayeri, Rouz, 47
Johnson, Ralph, 47

K
Katehi, Linda P.B., 84
Kelble, Jack R., 48
Kim, Sangtae, 91
Kincheloe, Iven C. "Kinch," Jr., 48
Klier, John, 89
Korsmeyer, Richard W., 49, 89
Kothari, Akshay D., 49
Krach, Keith J., 50
Kurtenbach, Aelred "Al" J., 50
Kyle, Richard G., 51

L
Ladisch, Michael, 89, 91
Lamberson, Donald L., 51
Lambertus, Peter K., 51
Landgrebe, David, 89, 91
Lane, Rita S., 52
Langster, Travis, 52
Leonards, Gerald A. "Jerry," 53, 91
Leonhardt, Fritz, 53, 89
Lievense, Jefferson, 90
Liggett, Emily M., 54
Linquist, Roger, 54
Lipps, Ben J., 54
Liston, Joseph, 55
Luckiesh, Matthew, 58
Lu, Luna, 55
Lundstrom, Mark S., 85, 90, 91

M
Mallof, Joseph T., 58
Malshe, Ajay, 92
Manian, Bala S., 58
Martin, John Joseph, 59
Martinson, John, 59
Massengill, Matthew E. "Matt," 60
McCulley, Michael J., 60
McDonald, John, 90
McEnroe, Paul, 61
McNett, Christopher Paul "Chris," 61
Mead, Elwood, 62
Melkote, Keerti G., 62
Michael, Harold, 90, 92
Miessner, Benjamin F., 63
Millard, Lindsay, 63
Miller, Jim, 63
Morgan, Burton D., 64
Morse, Stephen, 90
Moses, Beth, 64
Muasher, Marwan Jamil, 65
Mueller, George, 90

O
Oliphant, Elmer Quillen "Ollie," 65
O'Neil, William J. "Bill," 66
Owen, Douglas, 90

P
Pipes, Byron, 92
Poppe, Patricia K. "Patti," 66
Potter, Audrey A., 85
Powers, Audrey, 67
Powers, Mamon M., Jr., 67
Purcell, Edward M., 68
Pyrak-Nolte, Laura, 92

R
Raman, Arvind, vii, 85
Ramkrishna, Doraiswami, 92
Reklaitis, Gintaras, 92
Robeson, Lloyd, 90
Ross, David E., 68
Ruan, Xiulin, 69
Rumsey, Jennifer Weerts, 69

S
Sachdev, Anil, 90
Sage, Andrew, 90
Salvendy, Gavriel, 92
Sampson, Henry T., Jr., 72
Sanger, Kenneth, 72
Schuhmann, Reinhardt, Jr., 92
Schwartz, Richard J., 84
Schweitzer, Edmund O. "Ed," III, 72
Shah, Manahar R. "Manu," 73
Shalaev, Vladimir, 92
Sharaf, Essam Abdel-Aziz, 73
Sheridan, Thomas, 90
Siirola, Jeffrey, 92
Sinha, Kumares, 92
Skeen, Tonya M., 74
Slayter, R. Games, 74
Solberg, James, 92
Sözen, Mete, 92
Stamm, David A., 75
Stanney, Kay, 90
Stevens, Martha Dick, 75
Sutherland, John, 92

T
Taylor, Richard W. "Dick," 76
Terrell, Karenann "Kat," 76
Thapar, Hemant, 90
Thomas, Frank, Jr., 77
Thompson, Don, 77
Turpin, J. Clifford "Turp," 78

U
Ufford, Donald Allen "Don," 78
Underhill, Kimberly, 78

V
Viskanta, Raymond, 90, 92

W
Wang, Patrick S., 79
Webb, Josephine, 79
Weiner, Andrew, 92
White, Julie Kramer, 80
Woodall, Jerry M., 80, 92

Y
Yang, Henry T., 84
Yih, Yuehwern, 92

Z
Zakaria, Adel A., 81
Zucrow, Maurice J., 81

CREDITS

The following individuals assisted in the creation of this publication:

Laura Abbott
Christine Babick
Jeannine Bergeron
Kat Braz
Brittany Clements
Veronica Gonzalez
Will Gwin
Ashton Helmke
Charles Jischke
Devynn May
Andrea McIntyre
Sarah-Jane McNalley
Michele Miller
Saad Mukhtar
Della Pacheco
Heather Pflug
Joseph Pekny
Paul Sadler
Chelsey Grassfield Stone
Kate Theobald
Korina Wilbert

With special thanks to the Virginia Kelly Karnes Archives and Special Collections Research Center in the Purdue University Libraries and School of Information Studies.

ABOUT THE AUTHORS

Arvind Raman is the John A. Edwardson Dean of the College of Engineering and the Robert V. Adams Professor of Mechanical Engineering at Purdue University. He is internationally recognized for his research on vibrations and nonlinear dynamics in nanotechnology, biomechanics, and manufacturing. He has been instrumental in founding the Shah Family Global Innovation Lab in the College of Engineering, which fosters technology development and translation of technologies for sustainable development. Raman's notable contributions include advancements with the Atomic Force Microscope, an indispensable tool for scientists and the industrial community to better image and measure properties of complex materials. Additionally, simulation tools developed by Raman's group on the nanoHUB have been used by thousands of researchers worldwide. An ASME fellow and recipient of the ASME Gustus Larson Memorial Award, a Keeley Fellowship, a College of Engineering outstanding young investigator prize, and an NSF CAREER award, Raman has demonstrated a lifelong dedication to engineering innovation.

Jim Small is a lifelong storyteller whose career spans global marketing, media, and higher education. From launching the television series *NBA Inside Stuff* to working as the associate vice president for engagement at the University of Notre Dame, Small has played a role in creating influential iconic brands and record-setting marketing campaigns. His passion remains the same: crafting stories that move people to think, feel, and do.